D1740391

Wireless Networks

Series editor

Xuemin Sherman Shen
University of Waterloo, Waterloo, Ontario, Canada

More information about this series at http://www.springer.com/series/14180

Basile L. Agba • Fabien Sacuto • Minh Au
Fabrice Labeau • François Gagnon

Wireless Communications for Power Substations: RF Characterization and Modeling

 Springer

Basile L. Agba
Hydro-Québec's Research Institute
Varennes, QC, Canada

Minh Au
Hydro-Québec's Research Institute
Varennes, QC, Canada

François Gagnon
École de Technologie Supérieure
Montreal, QC, Canada

Fabien Sacuto
McGill University
Montreal, QC, Canada

Fabrice Labeau
McGill University
Montreal, QC, Canada

ISSN 2366-1186   ISSN 2366-1445   (electronic)
Wireless Networks
ISBN 978-3-319-91327-8   ISBN 978-3-319-91328-5   (eBook)
https://doi.org/10.1007/978-3-319-91328-5

Library of Congress Control Number: 2018942024

© Springer International Publishing AG, part of Springer Nature 2019
This work is subject to copyright. All rights are reserved by the Publisher, whether the whole or part of the material is concerned, specifically the rights of translation, reprinting, reuse of illustrations, recitation, broadcasting, reproduction on microfilms or in any other physical way, and transmission or information storage and retrieval, electronic adaptation, computer software, or by similar or dissimilar methodology now known or hereafter developed.
The use of general descriptive names, registered names, trademarks, service marks, etc. in this publication does not imply, even in the absence of a specific statement, that such names are exempt from the relevant protective laws and regulations and therefore free for general use.
The publisher, the authors and the editors are safe to assume that the advice and information in this book are believed to be true and accurate at the date of publication. Neither the publisher nor the authors or the editors give a warranty, express or implied, with respect to the material contained herein or for any errors or omissions that may have been made. The publisher remains neutral with regard to jurisdictional claims in published maps and institutional affiliations.

Printed on acid-free paper

This Springer imprint is published by the registered company Springer International Publishing AG part of Springer Nature.
The registered company address is: Gewerbestrasse 11, 6330 Cham, Switzerland

To my mother, Kinam, for guiding my first steps towards curiosity; To my wife, Clotilde, and my daughters, Dora and Eurielle, for their love!

Basile L. Agba

To my parents, Anny and Romain, and my sister, Aurore, for all their love and support.

Fabien Sacuto

Dedicated to my mother, Loan, my relatives and friends who have endured my lack of attention; To those who inspire me.

Minh Au

To Jehanne, Marine, and Amélie for their unwavering love and support.

Fabrice Labeau

To my family, which also includes all who want to contribute to our well-being through creating and sharing knowledge.

François Gagnon

Foreword

The electric industry experienced major transformations driven by the imperative of the energy transition and environmental constraints in the context of climate change. At the heart of these transformations, the advent of digital technologies will support an ever-growing flow of data and information in an increasingly connected world. The evolution of the substations, nerve centers that operate and manage the power grid, does not escape from this reality. To address these challenges, communication technologies including wireless technologies are an essential lever. The book you have in hands is part of this context.

As managers in charge of innovation, we are on the lookout for cutting-edge innovative technologies that improve significantly reliability, efficiency, and resiliency of the electric grid. It is therefore an honor to introduce the authors of this book on the frontier of electrical engineering and telecommunications. The authors are scientists with extensive experience in these aforementioned scientific disciplines. Indeed, among the authors, you will find:

- Industrial Chairs holders who have demonstrated their commitment over the past years, during which their research has resolved major operational and practical problems. The many awards they have received from the scientific community and industrial committees reflect the importance of their contributions.
- Researchers and engineers who combine several years of practice in industry, specifically in the electrical industry, provide a pragmatic vision that takes into account the specific criticality and reliability that are required to operate in the electricity grid. Through peer recognition of their contributions, they have established a positive influence on their discipline.

This book is also a successful example of the synergy between academia and industry that we wish to personally acknowledge.

Throughout the book, you will find what has been achieved over the last 5 years. We personally convinced that the authors are featured prominently for writing this book. The central topic of the book is the deployment of wireless technologies in environments with strong impulsive noise disturbance. This problem is even more important considering the rapid and the increasing demands of sensors and intelligent electronic devices in these environments. Wireless sensor networks offer significant benefits that will serve, on the one hand, to inquire the current condition of the equipment for purposes of continuity and preventive maintenance and, on the other hand, to monitor and control them remotely. Besides an excellent review of the state of the art, the book covers two essential aspects:

- How to conduct a safe measurement campaign in a high-voltage environment and what are the appropriate radio frequency characterization techniques.
- How to derive analytical models from the characterization as well as quantifiable parameters to use for different substation voltage level.

We strongly recommend this book for practitioners of substations automation and more generally to those who wish or are willing to deploy wireless systems in industrial environments highly disturbed by electromagnetic interference. They will find valuable tools and tips to optimize the design of such systems for achieving high performance and reducing the risk of malfunction. In this book, the authors propose analytical models that can emulate these harsh and hostile environments. These can in turn be used for the deployment of new applications without the resources and time constraints of measurement campaigns and characterization. Finally, scientists, students, and simply curious readers will leverage their knowledge on both electrical engineering and telecommunications.

Concise and easy to read, at the end the issue of the wireless infrastructure deployment in substations and the proposed solutions will have no secrets for you.

Enjoy your reading!

Principal Director   Jean Matte
Institut de recherche d'Hydro-Québec
Varennes, QC, Canada
February 2018

President   Réal Laporte
Hydro-Québec - Innovation, équipement et services partagés
Montreal, QC, Canada

Preface

In the context of the next generation of electric grids, called smart grids, the use of modern and advanced communication systems in high-voltage substations can significantly improve the efficiency, reliability, and safety of the electric power grid. Indeed, the deployment of intelligent sensor networks allows for the development of a more efficient, rapid, and automated remote monitoring, control, and diagnosis in major pieces of high-voltage equipment in substations: current, potential, and power transformers, circuit breakers, and high-voltage disconnectors. Workers and maintenance supervisors on high-voltage substation sites require more information from major high-voltage (HV) equipment, such as power transformers and circuit breakers, to be collected via a network of electronic intelligent devices. However, using wired sensor networks might be complex to manage in terms of wiring complexity, cost reduction, and ease of deployment. Wireless sensor networks offer significant benefits in this area. Unfortunately, high-voltage substations are harsh and hostile environments whose wireless communication systems can be interfered to such an extent as to render their performances severely degraded.

This book aims to provide a general overview of electromagnetic interferences (EMI) and describe the measurement and characterization methods associated with them, in keeping with the latest research. Impulsive noise phenomenon in substations is particularly emphasized. A review of related literature shows the impact of impulsive EMI sources on wireless communications; namely, that performance can be severely degraded in these special environments. Through the literature review, existing (partial discharge) PD characterization methods do not assess if electromagnetic radiation from PD activity is a source of interference for the radio communication systems operating in the industrial, scientific, and medical (ISM) radio bands and existing impulsive noise models cannot link the physical characteristics of high-voltage installations to the induced radio interference spectrum neither.

Access to power substations is very limited, even for the staff of electricity providers. Despite the access limitation, we have managed to measure RF noise in several substations and we have recorded around 120 sequences of noise samples. The work we propose does not only provide information regarding RF noise within substations, but it also provides the means to replicate such a noise using different statistical models. We present classic statistical models for impulsive noise and our own models. In addition to the noise modeling, we also provide parameters that can be used to replicate a representative impulsive noise characterizing a substation working under a specific voltage. Our goal is to provide the means for anyone who wants to simulate a wireless communication within power substation to be able to do so without having to perform any measurements; the only information the reader would require is the power equipment voltage and the wireless technology he wants to test.

This book proposes new characterization methods of EMI phenomena in substations. This aims to provide a noninvasive measurement and detection method for the characterization of the electromagnetic radiations induced by PDs. This information can then be applied towards the development of rapid online remote monitoring and diagnostic tools in (high-voltage) HV equipment, and/or for characterizing and modeling wireless channels in substations. These impulsive interferences are measured using a wideband antenna surrounded by HV equipment in normal operation. This is a complete and coherent approach that links physical characteristics of PDs to the induced radio interference spectrum and may be used in the design of new remote control and monitoring systems via wireless IED (WIED), as well as for the performance analysis of wireless communications.

The book is aimed at readers working in electrical engineering, power engineering, telecommunications, and automation. The reader profile is mostly industrial, but readers from academia may also be interested, since research in smart grid is becoming more and more popular among electrical, computer, and mechanical engineering departments in universities.

The primary audience of this monograph are the electrical and power engineering industry, electricity providers, and all companies who are interested in substation automation systems using wireless communication technologies for smart grid applications. However, high-voltage substations are considered harsh and hostile environments for wireless communication systems due to impulsive interferences. This book aims to give an overview of most recent research on characterization and channel modeling in power substations. These results may be used in the design of new remote control and monitoring systems, as well as for the performance analysis of wireless communications. This book will provide recent contributions to the deployment of wireless communication systems in substations where significant improvement in protection, control, automation, and monitoring applications in high-voltage equipment can be achieved.

Academic researchers will also be interested in this topic because wireless communications within power substations are increasingly becoming the subject of collaborations between industries and universities for the smart grid. Researchers and engineers with a wireless communication background (industrial or academic) will also be interested in this work. In particular, for wireless or other communication systems that are impaired by such a harsh and hostile impulsive interferences. This work is not only limited to substation environments. It is intended for researchers and engineers who aim to design devices that need to deal with impulsive jammers, lightning discharges, and impulsive noises in Power Line Communications (PLC). Furthermore, partial discharges in substations have been observed on electrical insulations. They cause irreversible damage and possible failure of high-voltage equipments. This work can provide signal processing algorithms for fast PD identification, localization by using low-cost wireless intelligent devices.

As far as we know, the topic of our work has not been rigorously investigated or published in a book. The main reason for the lack of publications is that substations are very difficult to access and most of the work related to partial discharges, Corona effect, or impulsive noise is limited to its own topic. There is a correlation between power equipment, electrical operations, partial discharges, Corona effect, and impulsive noise that is difficult to highlight due to the involvement of different backgrounds in electrical engineering and research domains such as power engineering, signal processing, digital communications, and statistical modeling. Most of the reference literature deals separately either with partial discharges in power substations or impulsive noise for communications. There is no major work explaining the correlation between RF impulsive noise in power substation and its electrical source: the partial discharges and the Corona effect that are both dependent on the power equipment (voltage and insulator nature).

Montreal, QC, Canada   Basile L. Agba
February 2018   Fabien Sacuto
                                                                          Minh Au
                                                                 Fabrice Labeau
                                                             Francois Gagnon

Acknowledgments

We would like to take this opportunity to express our sincere gratitude to several distinguished colleagues and partners-in-mind from the academia, the world of business, and scientific research organizations who have rendered their valuable assistance for the completion of this book. Most specifically, our gratitude goes to the following: Hydro-Québec and their supportive employees whose fruitful discussions and constructive feedback have significantly improved this book. A special thanks goes to Mélanie Levèsque, Calogero Guddemi, Jean Béland, Sylvain Morin, Marthe Kassouf, Ryszard Pater, and Sébastien Poirier.

The authors acknowledge the financial support of the Natural Sciences and Engineering Research Council of Canada (NSERC) and MITACS. They are also grateful to Susan Lagerstrom-Fife from Springer for her support throughout the entire production process.

About the Authors

Basile L. Agba (IEEE Senior Member—2013) is a senior scientist and project leader at Hydro-Quebec Research Institute (IREQ). He holds an M.Sc. and a Ph.D. in electronics and optoelectronics (University of Limoges, France). His main research interests include channel modeling in high-voltage environments, fixed terrestrial links design, wireless systems, and RF design. He has a special interest in smart grid with focus on communications systems, cybersecurity, and substation automation. He has authored more than 60 refereed papers in refereed journals and conference proceedings in these areas. Since 2009, Dr. Agba is an adjunct professor with the Electrical Engineering Department, ÉTS (École de technologie supérieure, Montreal-Canada). He is also an active member of International Telecommunication Union, Study Group 3 on radio propagation.

Fabien Sacuto holds a Bachelor in Electrical and Computer Engineering from the ESME-Sudria engineering school, Ivry-sur-Seine, France. He received his Master of Engineering in Telecommunication Networks from the École de Technologie Supérieure (ÉTS), Montreal, Canada, in 2010, and he received his Ph.D. in electrical and computer engineering from McGill University, Montreal, Canada, in 2016. Dr. Sacuto specializes in wireless communications, RF noise measurements and modeling, and most specifically impulsive noise from high-voltage environments. He has been working on wireless communications within power substations with Hydro-Québec from 2009 to 2016 as external consultant, and as postdoctoral researcher. He has been involved in smart grid projects with Hydro-Québec on substation automations, sensor synchronizations, and wireless systems. He is currently working as a research professional at ÉTS on low latency communications, QoS estimation for VoIP, and the Internet of Things (IoT).

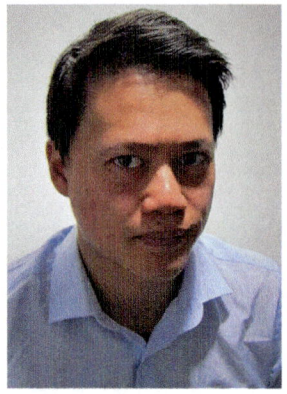

Minh Au is a researcher at Hydro-Québec Research Institute (IREQ). He holds a B.S. and M.S. degrees in computer science and telecommunication from the University of Poitiers, France, and a Ph.D degree from École de technologie supérieure, Montreal, QC, Canada. He has been actively involved in the Richard J. Marceau Industrial Research Chair for Wireless Internet in developing countries with Media5 as a postdoctoral fellow. He serves as an expert reviewer for several conferences and journals of IEEE and IET. His current research interests include channel modeling in substation, partial discharge phenomenon, low-latency communication for machine-to-machine, information theory on coding theory, and cyber-physical security in smart grid.

Fabrice Labeau is the Associate Dean (Faculty Affairs) in the Faculty of Engineering of McGill University, where he holds the NSERC/Hydro-Québec Industrial Research Chair in Interactive Information Infrastructure for the Power Grid. His research interests are in applications of signal processing to healthcare, power grids, communications and signal compression. He has authored more than 175 refereed papers in refereed journals and conference proceedings in these areas. He is the director for operations of STARaCom, an interuniversity research center grouping more than 50 professors and 500 researchers from 10 universities in the province of Quebec, Canada. He is senior past president of the Institute of Electrical and Electronics Engineers (IEEE) Vehicular Technology Society, President of the IEEE Sensors Council, and the past chair of the Montreal IEEE Section. He was a recipient in 2015 and 2017 of the McGill University Equity and Community Building Award (team category), of the 2008 and 2016 Outstanding Service Award from the IEEE Vehicular Technology Society, and of the 2017 W.S. Read Outstanding Service Award from IEEE Canada.

Franois Gagnon holds a B.Ing. and Ph.D. from the École Polytechnique de Montréal, and has been a professor at the École de technologie Supérieure (ÉTS) since 1991. He has held industrial research chairs since 2001. In addition to holding the Richard J. Marceau Industrial Research Chair for Wireless Internet in developing countries with Media5, François Gagnon also holds the NSERC-Ultra Electronics Chair in Wireless Emergency and Tactical Communication. He also founded the Communications and Microelectronic Integration Laboratory (LACIME) and was its first director. He has been very involved in the creation of the new generation of high-capacity line-of-sight military radios offered by the Canadian Marconi Corporation, which is now Ultra Electronics TCS. Ultra-Electronics TCS and ÉTS have obtained the NSERC Synergy prize and an ADRIQ partnership prize for this collaboration. He is actively involved in the SmartLand project of UTPL, Ecuador, the STARACOM strategic research network, and recently with the Réseau Québec Maritime.

Contents

Chapter 1
Introduction

Two of the most pressing contemporary environmental issues are climate change and the management of scarce energy resources. The increased awareness of these issues encourages the electricity industry to deploy advanced communication technologies within the electric grid for the next generation of electric grids, called smart grids [1–3]. Since the early twenty-first century, this novel concept has emerged to manage the increasing demand for energy resources while maintaining economic growth and environmental sustainability. This new paradigm allows for decentralized production by integrating renewable energy resources and enhancing consumer empowerment for sustainable development [4–6].

Over the last few years, the development of smart grid has become a promising area of research for both industrial and academic researchers. We mention the development of:

- new information and communication technologies for better electricity management;
- battery energy storage systems for improving the power quality, reliability and energy efficiency of the electric grid;
- smart grid interoperability for the integration of renewable energy resources.

Readers interested in knowing more about the Potential applications and benefits of the smart grid have been summarized and are referred by [3, 4, 7, 8].

1.1 Motivation

Electrical providers have started studying the possibility of integrating telecommunications to the grid to optimize energy distribution according to the consumers' needs [9]. The cohabitation of telecommunications and electricity network is an essential component of the so-called *Smart Grid*. One of the possible actions of the

© Springer International Publishing AG, part of Springer Nature 2019
B. L. Agba et al., *Wireless Communications for Power Substations:*
RF Characterization and Modeling, Wireless Networks,
https://doi.org/10.1007/978-3-319-91328-5_1

smart grid would be to collect and exploit information about energy consumption of homes in order to distribute the appropriate amount of energy required by the consumers' devices. This vision of new electricity distribution requires the installation of communication networks in homes, but also in substations. The International Electrotechnical Commission (IEC) has created a standard, IEC 61850, in order to manage substation automation (SA) [10]; SA encompasses monitoring and control of substation equipment by using networks of Intelligent Electronic Devices (IEDs) that are connected to power equipment. Communication systems aim at exchanging information between transmitters and receivers with specific throughput and reliability; moreover, both hardware requirements and communication performances influence the selection of the communication technology for a specific application.

The choice of a communication technology for IED networks in substations depends on both the environment and applications. Many communication technologies offer possibilities to install IED networks interacting with substation equipment. Before going any further, we summarize the communication media that might be considered for SA:

- **Wireless communications**: they include Wi-Fi, Wimax, LTE [11–13], Bluetooth, ZigBee and cellular. Wireless technologies allow an easy installation without requiring a shutdown of the substation. The main issues are the lifetime of the battery of equipment if it is not attached to a reliable power source, security of the data and a communication channel that is more susceptible to electromagnetic noise and interferences.
- **Optical fiber**: this technology is immune to electromagnetic noise and it offers very high communication throughput. Optical fiber can be used for IED networks but the installation cost (on the order of $1000 per meter including digging) remains a drawback. Ideally, new substations can integrate optical fiber into their structures. However, for existing substations, the installation of optical fiber would require extra retrofitting, which might become even more expensive.
- **Power line communications** (**PLC**): this method uses the existing power lines to communicate between IEDs. This technology is also affected by electromagnetic noise coming from equipment radiation and from other sources active in the wires.

The installation of telecommunication systems, such as wireless and PLC communication systems in substations would allow to interact with, control and monitor the equipment from the outside of the substation site.

Among these three aforementioned communication technologies, wireless systems are strong candidate that allow easy installation of IEDs and require no retrofitting of substations that would have been very expensive [7, 14–17]. Installing wireless IEDs in environments such as substations encounters issues that are common to wireless communications: interference and noise. The major problem in substations is that power equipment generates "impulsive noise" [18–22] in a band covering the carrier frequencies of classic wireless communications. Substation impulsive noise is very different from classic white Gaussian noise, as the noise is correlated [20, 21, 23] and the samples have reported to follow a Gaussian mixture distribution [24–26].

To make wireless networks of IEDs effective in substations, the receiver must possess information about the impulsive noise. Information about impulsive noise in substation includes the pattern of the noise and the dependence of its characteristics with the substation environment. Measurements and statistical modeling of impulsive noise are essential to the design of optimum receivers that are capable of mitigating the impact of substation noise on the communication.

In this monograph, we particularly focused on electro-magnetic interferences (EMI) induced by discharges in high-voltage (HV) installations. Partial discharges (PD) that take place in the air are the predominant source of EMI. Their activity can take place in HV equipment: along the insulation surface or in an air gap located between two separated conductors. When the local electric field is sufficiently high, the charged particles can collide, which produces an electron avalanche process in which electric discharges can occur. The induced charges and currents generate high impulsive electromagnetic radiations also observable as "impulsive noise".

1.2 Monograph Organization

This monograph is organized to show the progression of our work including the gathering of information about power substations, the measuring RF noise generated by power equipment and electrical operations, and finally the providing of realistic models representing RF environment of substations.

We introduce wireless communications and substation environments in Chap. 2. The purpose is to provide a background on existing wireless technologies and substation noise and highlight the fact that many existing wireless communication systems do not exploit information about substation environments. The background also covers existing attempts at installing wireless sensors in high voltage environments. We also present in this chapter the nature of substation impulsive noise, and in particular its two main sources: partial discharges and Corona effect.

In order to observe the RF environment of power substations, we describe in Chap. 3 the measurement setup that we have used to measure sequences of sampled impulsive noise in different substations. From this measurement campaign, we provide impulsive noise characteristics for different substation voltage ranges, which also contributes to the computation of representative parameters for existing impulsive noise models.

The measurement campaign provides all the information required for designing realistic models of power substation RF environment. We propose first, in Chap. 4, a coherent, detailed and validated PD model that links the physical characteristics of HV installations to the induced radio interference spectrum. PD sources are generated experimentally in a 16 kV stator bar in laboratory. Using our proposed characterization process described in Chap. 3, our proposed model is validated by comparing measurement results to the results of simulations.

We propose, thereafter in Chap. 5, a general RF impulsive noise waveform model based on second-order statistics. Time series models are used to estimate spectrum characteristics of RF impulsive signals from PDs with a reasonable number of parameters. It is shown that residuals from fitted time series models are stochastic processes in which the variance is not constant over time. A heteroskedastic white Gaussian noise is used to reproduce the random behaviour of transient impulsive waveforms. Measurement results and simulation results show that our RF impulsive noise waveform model fits measurements accurately.

A statistical model that requires only observations of the environment can be very useful for implementing robust wireless receivers in ISM (Industrial, Scientific and Medical) band; therefore we present, in Chap. 6, a novel statistical noise model in wide band. The model uses a partitioned Markov chain (PMC) in which each state of the chain is associated with a Gaussian distribution. We show how the states are connected with each other and we provide parameters estimated from the measurement campaign by using a novel impulse detection method. To complete the RF representation of power substations, we remind power substations host multiple sources of noise.

In Chap. 7, the RF impulsive noise model is generalized in the presence of multiple PD sources. A spatial and temporal Poisson point process (the Poisson field of interferers) is used to emulate an environment typical of substations. This generalized model can take into account physical-statistical parameters estimated from data. We employ our proposed RF impulsive noise waveform model to reproduce typical impulsive noise waveforms in which parameters are estimated from the measurements that were presented in Chap. 5. Moreover, the first-order characteristics of PD such as power density, inter-arrival time and occurrence distributions from data can be used. In this chapter, the statistical properties of the generalized model can be derived based on practical assumptions. Therefore, signal processing algorithms can be implemented for rapid PD identification, localization, and impulsive noise mitigation techniques. These can in turn be utilized in wireless communications in substations.

1.3 Contributions

In this monograph, the study of wireless communications in substation environments has helped to highlight and to analyze the correlation present in the impulsive noise process. The design of a novel correlated noise model allows us to generate synthetic noise that is more similar to the measurements than other existing models. The implementation of the proposed noise model in a maximum a posteriori (MAP) receiver allows us to mitigate the impact of impulsive noise. The measurement campaign provides an important amount of data that can be used to study impulsive noise, to compare the different noise models and also to verify the accuracy of the models. The measurements of substation impulsive noise can also help to verify the performances of receivers that are designed to mitigate the impact of substation noise.

We summarize below the major contributions of the work described in this monograph:

- We have studied impulsive noise characteristics and the correlation between the noise samples that create the impulse pattern. From our measurement campaign, we have gathered enough noise samples to provide accurate information about impulsive noise in substations, such as the power spectrum of the impulses, the correlation between the impulse samples, the correlation between the environment's voltage range and the impulsive noise characteristics and the possible correlation between the impulse generation and the frequency of the equipment voltage (60 Hz).
- We have introduced a new method to detect and isolate impulsive sections in the noise. A reliable way of calculating parameters for impulsive noise models is to separate the impulses from the background noise. The impulse detection method that we propose works in two steps. The first step consists in evaluating the background noise level. The second step aims at detecting the samples that belong to impulses based on a rule that compares the sample values with the background noise level. With this method, we are able to calculate very accurately parameters for our new PMC model, but also for existing impulsive noise models.
- We have proposed a full characterization of EMIs produced by PDs and generic impulsive noise models for wireless channels in substations. The proposed model is a complete and coherent approach that links physical characteristics of PDs to the induced radio interference spectrum. It allows for the performance analysis of wireless communication systems as well as the design of robust receivers corrupted by impulsive noises in substations. This work is based on three major axes of research, namely: (a) the characterization of radio frequency (RF) signals from PD activity in HV equipment; (b) the formulation of a generic RF impulsive noise model induced by PD; (c) the statistical analysis of the model in the presence of multiple PD sources.
- We have built a model of wide band substation impulsive noise. Using a state diagram, we have designed a model that can replicate impulsive noise samples with the appropriate correlation for a wide band representation. The impulses are generated with a damped oscillating waveform, which is similar to the impulse pattern observed in measurements. The model is called Partitioned Markov Chain (PMC) and requires between 2 and 19 states. The number of states depends on the degree of accuracy of the representation; the more states are used, the more accurate the model is. Our PMC model can be used to represent impulsive noise from any kind of power substation for the 780 MHz to 2.5 GHz band.
- We have proposed a new parameter estimation method for our PMC model. This parameter estimation method uses a Fuzzy C-means (FCM) algorithm and it provides better results than our first method for replicating impulsive noise characteristics in terms of sample values, impulse duration and repetition rate. Once the PMC model is estimated with this method, it generates impulses with power spectrum much more similar to the measured impulses than existing models.

- We have provided representative noise characteristics, such as impulse duration, repetition rate and impulse amplitude, and we have calculated parameters of impulsive noise for different substations and for different noise models, including PMC. The parameters are provided for different voltage ranges in substation, which allows to predict the behavior of the noise depending on the voltage of the equipment in the vicinity of the communication system. Anyone using the PMC model with the representative parameters can generate a realistic impulsive noise in wide band, without having to go and measure the noise in existing substations.
- We have implemented a less complex configuration of PMC, called PMC-6, in a MAP receiver as a proof of concept that our model can help mitigating substation impulsive noise. The receiver exploits the memory of the model to calculate the probabilities useful to the detector, the de-mapper and the decoder in order to estimate the most probable transmitted bit based on a received frame. The use of the model memory enhances the performance of the receiver and mitigates the impact of impulsive noise better than receivers expecting an uncorrelated noise.

Chapter 2
EMI and Wireless Communications in Power Substations

2.1 Introduction

This chapter gives a general overview of EMI and describes the measurement and characterization methods associated with them, in keeping with the latest research. Impulsive noise phenomenon in substations is particularly focused and existing impulsive noise models are discussed. The chapter is organized as follows: Sect. 2.2 introduces, defines, and classifies the concept of EMI sources. A review of related literature shows the impact of impulsive EMI sources on wireless communications; namely, that performances can be severely degraded in these special environments. In Sect. 2.3, the emphasis is on how EMIs are generated mainly by partial discharge sources in substations. The physical mechanisms, measurements, detection methods and existing PD models are reviewed. In Sect. 2.4, the characterization and existing impulsive noise models for communication channels are presented. Section 2.5 wireless communication technologies that have been used in substations are reviewed. Section 2.6 concludes this chapter with a brief summary.

2.2 Concept of EMI and Classification

This section presents a general overview of EMI, which can be classified into two categories: natural noise sources and man-made noise sources. In high-voltage substations, man-made noise sources are caused mainly by electric arc discharges in which electromagnetic radiations are sources of interference for conventional wireless communication systems.

© Springer International Publishing AG, part of Springer Nature 2019
B. L. Agba et al., *Wireless Communications for Power Substations:
RF Characterization and Modeling*, Wireless Networks,
https://doi.org/10.1007/978-3-319-91328-5_2

2.2.1   Definition of EMI Sources

Electromagnetic interferences are observed when external sources interfere with any electronic devices or with communicating systems. Their tolerance to unwanted electromagnetic (EM) radiation should be analyzed and characterized in order to assess the electromagnetic compatibility (EMC) of interference sources [27, 28].

We may distinguish intrinsic noise sources inherently generated by the electronic device itself and the external noise generated by interferences. Intrinsic noise sources are generated inside the device. A typical example of intrinsic noise is thermal noise. It comes from the motion of free electrons inside a conductive material, which is inherently random and unavoidable. The resulting signal amplitude fluctuates randomly. This can be modeled as Gaussian noise in which the mean value is zero and its variance is given by:

$$v_n^2 = 4k_B T R \Delta f \tag{2.1}$$

where k_B is Boltzmann's constant, T is the temperature, R is the resistance of the material and Δf is the bandwidth of the measurement system that has been employed.

An external noise is an unwanted signal whose EM waves come from undesired sources. In the literature [27, 28], these interferences are classified into natural noise sources (e.g. atmospheric noise, cosmic noise, etc.) or man-made noise sources (e.g. industrial noise, arc welding, switches, etc.).

2.2.2   Natural Noise Sources

These noise sources are produced by charged particles that are either present or produced naturally in the environment. The waveform can be modeled as Gaussian noise (e.g. cosmic noise) or as transient impulsive noise (e.g. lightning discharges). The emitted radiations can occupy a wide range of frequencies; typically, they have a bandwidth of a few GHz.

Atmospheric noises are generated by lightning discharges in thunderstorms. They are characterized by fast transient waveforms and rapid time decay (10–100 μs). The current amplitude can be 20 kA for an average arc discharge and 300 kA for a very high-arc discharge. They are mostly harmful to humans and damage electronic devices embedded in aircraft. The frequency spectrum can span to 20 MHz and the spectral density exhibits a typical $\sim 1/f$ decrease with frequency [28]. This spectrum content explains why natural noise sources are generally considered less troublesome than man-made noise sources, which cover a much broader band.

2.2.3 Man-Made Noise Sources

Electromagnetic man-made noise sources are generated by human activities via any electrical or electronic devices that may be located in industrial, residential or business areas. These sources can generate various types of waveform like stationary random signals, non-stationary random signals (e.g. impulsive noise), and/or modulated signals from undesired communication systems.

Interference sources can affect the reliability of communication systems and also damage electronic devices when the amplitude of the current is very high. These sources can be intentional interferences (jammers) and/or unintentional interferences. The latter could come from EM sources induced by the normal operation of any electronic device. As a result, it is necessary to characterize and to quantify their impact on communication systems.

Among man-made noise sources, impulsive noise can be generated by automotive ignition systems, arc welding, power-line distribution systems, high-voltage transmission lines or HV equipment in substations. The amplitude associated with such sources is very high and has short durations compared to background noise: 50–310 μs in power-line distribution systems [23, 29], and 50–200 ns in HV transmission lines as reported in [30, 31]. The impulse repetition rate can range from 50 to 200 kHz in power-line distributions and around 50–100 kHz in HV transmission lines. When impulsive noise is significant, communication performances can be severely degraded.

2.3 Electromagnetic Interference in Substations

2.3.1 Functions of Power Substations

Power substations are built to either transport or distribute electricity from energy sources (dam, gas station, eolian) to the consumers by using specific power equipment. Generally, substations are situated outdoors, with an area of 1 km × 1 km on average and electricity is transported through power lines and buses (Fig. 2.1) toward other pieces of equipment to perform electrical operations [32, 33]. We refer to the electrical procedure resulting in distributing a certain amount of energy to different locations of the substation and ultimately to the downstream grid as electrical operations.

The size and the structure of the equipment might affect the propagation of wireless transmissions by creating reflections, losses and obstructions. Those issues need to be considered for wireless installations, however, in this monograph work, we only consider a direct transmission between two communications points, respectively the transmitter and the receiver, where the receiver is disturbed by surrounding impulsive noise. Several wireless technologies could be used for the transmission; we consider a wide band study of substation Radio Frequency (RF) environments in order to be able to select a specific carrier frequency of communication for future studies.

Fig. 2.1 735 kV area of a Hydro-Québec power substation (*Courtesy of Hydro-Québec, 2014*)

2.3.2 Pieces of Equipment and Electrical Operations

This section presents the different pieces of equipment which operate in substations. The location of the equipment is determined by the requirement of transporting the voltage in the direction of lower-voltage substations neighboring the energy consumers [33, pp. 93–95]. The equipment configuration and location must consider the transport distance to provide a sufficient power to reach the distribution substations. Substations with several voltage ranges are also configured to allow electrical operations aiming to transport a specific voltage from a location within the substation to another. The equipment working in power substations is described as follows:

- Voltage transformers aim to convert an input voltage to a different output voltage. The transformer coefficient is determined by the layout of a core with primary and secondary windings. The circuit (core and windings) is placed in a tank, usually made of aluminium, for the largest ones, with an oil-cooling system [34, pp. 9–12]. Bushings, mostly composed of porcelain, isolate the connections between the transformer and the external current bearing circuit across the substation. We will see later how the transformer inner portions and bushings are potential locations of impulsive noise sources.
- Current transformers are basically designed like voltage transformers except that the internal circuit is built to provide a secondary current proportional to the input current with a phase shift [33, pp. 157–160].
- Circuit breakers are boxes hosting several mechanical switches that let electricity circulate to another part of the substation depending on the maintenance requirements. Usually the circuit breaker operations happen on average once or twice a day and are used to shunt the voltage to other equipment.

- Bus bars are non-isolated conductors that transport high voltage between the electrical pieces of equipment inside the station [33, pp. 98–99].
- Power lines are similar to bus bars but for a larger scale of distances. The electrical cables are generally made from aluminium alloy reinforced with steel strands. They are located all around the substation to connect the substation equipment to the outside world. The lines are carried by 20-m high metallic towers [33, pp. 418, 683–685].
- Switch gears are switching devices associated with controlling, measuring and regulating of the equipment. These metallic structures interrupt the electric transmission between pieces of equipment, when the electric network requires it [33, p. 467]. The switch gear operation consists in separating two electrified metallic arms, which creates a long arc in the insulator for a few seconds.

All types of substation equipment are composed of conductors fed with high voltages and separated from each other by insulators of different nature. The impact of these components on the characteristics of the RF noise is described in the next section.

Corona or partial discharges are defined as localized electron transfers from a conductor to another, generated from a transient state of an insulator when the electric field emitted from a conductor exceeds a critical value [35]. However some differences exist between partial discharge and Corona effect in their generation and their observation. In this section, we introduce the electrical phenomena of partial discharge and Corona effect and we explain how they can be produced in air-insulated substations. We also describe a setup that can reproduce partial discharges in a laboratory setting in order to generate impulsive noise.

2.3.2.1 Corona Effect

The Corona effect corresponds to the ionization of a gaseous insulator around an electrical conductor [18, 35]. The electrified conductor drives an electric field in the radial direction, and when this field exceeds a critical value characterizing the gas, the neighboring electrons are ejected. The Corona effect is generally observable along power lines and it is an important source of impulsive noise [36–38].

During a measurement campaign reported in [18], Pakala observed another noise near power lines, similar to Corona effect, and called it Gap noise. The air electrons are driven by the electrical field coming from the lines and are attracted by metallic structure of the tower holding the lines [39]. The term "gap" comes from the distance between the metallic conductors, here the line and the tower (Fig. 2.2). The Corona effect and Gap noise are different kinds of noise but they have some similarities. Both come from a voltage that creates an electric field exceeding a critical value. When this critical value is reached, a discharge, complete or partial, occurs with physical manifestations like acoustic noise, sparks, electric arc, ozone smell, and UV radiations.

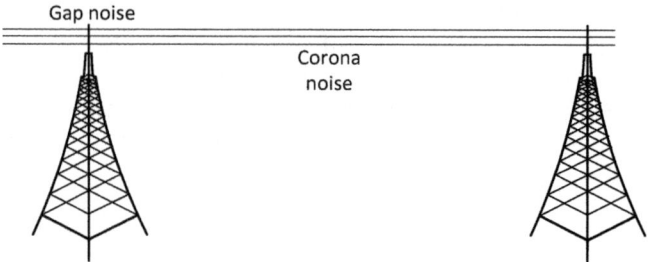

Fig. 2.2 Corona and gap noises near power lines

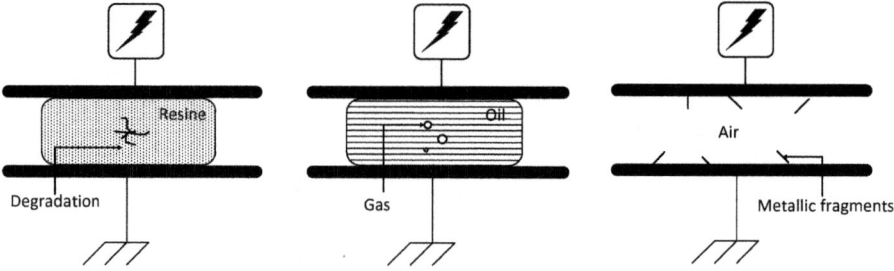

Fig. 2.3 Different types of partial discharge

2.3.2.2   Partial Discharges

The partial discharge is an electrical phenomenon that occurs inside an insulator or at the surface of an electric conductor. When an insulator is degraded due to aging, some microscopic areas may be less efficient than the rest of the insulator. In the case of two conductors under different electrical potentials and separated by such an insulator, an electrical field is emitted from a conductor to the other (Fig. 2.3). When this field is driven over a specific value of the microscopic degradation, a movement of ions creates a partial current [35]. This current occurs each time the critical value is reached by the electric field, which could happen several times in a half 60 Hz cycle. As shown in Sect. 2.2.1, a transformer contains an internal circuit cooled with oil. Although oil is an electrical insulator, it is usual that numerous gas bubbles, with a different dielectric strength, appear [34]. When those gas particles get closer to an electrified conductor, they host partial discharges, whose electromagnetic field can pass through the external circuit connections. Also the external bushing becomes degraded with aging and the porcelain imperfections can also host partial discharges [22]. In the same case of the Corona effect, partial discharges are observable as impulses with a very short rise time and a longer fall time. In comparison to an electrical discharge, in electromagnetic domain, this waveform seems to be "modulated" giving a damped oscillating shape to the impulses.

2.3.3 Early Impulsive Noise Measurements

Generally, power lines, circuit breakers and bus bars usually host Corona and Gap noises [18], whereas transformers host internal and external partial discharges [21, 22, 40, 41].

Based on the electrical nature of partial discharges, corona and gap noises, there are different ways to create impulsive noise in the laboratory. First, the laboratory must provide a high voltage source and a ground grid. Second, a partial discharge specimen with degraded insulator or conductor (Fig. 2.3) must be connected to the generator and the ground. Finally, to observe the impulsive noise, an electrical or RF setup can be installed to collect electromagnetic (EM) radiation while the generator delivers a voltage potential exceeding the specific value of the microscopic degradation within the insulator.

Previous measurement campaigns in air-insulated substations were first performed in the spectral domain [18, 19, 22]. Pakala created a measurement setup using antennas and spectrum analyzers and placed it in the vicinity of high voltage equipment inside the substation. By surrounding with antennas a piece of equipment, generally power lines, and by averaging their power spectrum, he followed the evolution of the noise for different voltages (from 80 to 800 kV). The bandwidth was first studied from DC to 1 GHz and then until 10 GHz. Pakala performed measurements under different weathers and for different line voltages. They observed that the noise power increases with the line voltage and also in rainy conditions [19]. The noise power was about -50 dBm for 900 MHz [18] and -60 dBm for 1–10 GHz [19].

With the improvement of measurement devices [42–44], measurement campaigns have now the ability to measure high voltage environments in the time domain. The measurements show that the impulsive noise is mainly composed of series of short damped oscillations, about 1 μs in duration [21], that depend on the feeding voltage of the equipment, the weather [22] and the nature of the insulators [20, 21]. The measurements in time domain show the time-correlated nature of the impulsive noise [45, 46] and also the correlation of the impulse occurrences with the 60 Hz feeding voltage [21]. Although the impulse generation is random, when an impulse occurs, the samples are generated with a damped oscillation waveform, which provides some time-correlation between the samples. A reliable representation of impulsive noise in substations must consider not only the generation of the impulses, but also the correlation between the samples that provides the damped oscillating waveform.

2.3.4 Ionization Process and Electrical Discharge in Gases

Discharges in gases are related to a partial or complete breakdown of a gas phenomenon. They occur when an applied electric field is sufficiently high. In such

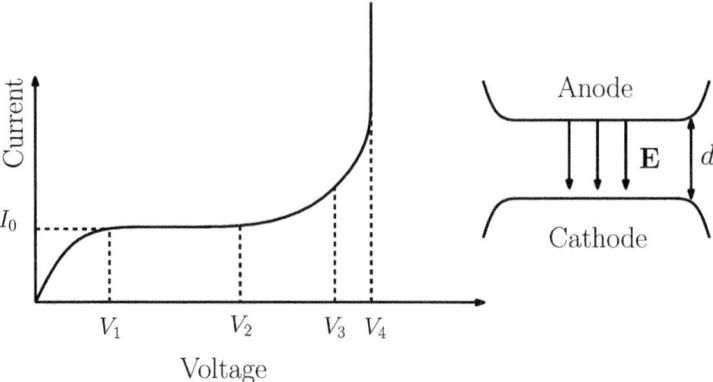

Fig. 2.4 Current–voltage relationship in prebreakdown region (*Courtesy of Kuffel et al.* [47])

instances, due to a strong acceleration of free electrons, other neutral molecules and atoms become excited or ionized by collisions in which kinetic energies are exchanged. An ionization process in gases can take place by avalanche effect. Depending on the nature of the gas, the ionization process can be significant when densities of electrons are high [35, 47, 48]. The study reported in [49] found that the current through a uniform field air gap, grows exponentially when the applied voltage is sufficiently high, as shown in Fig. 2.4. A pair of electrodes is separated by an air gap of a length d, and the electric field $\mathbf{E}$ between the electrodes is generated by an applied voltage. At voltages higher than V_2, the current growth is due to ionization by electron collision in the gas [49].

Various ionization phenomena can be observed during an electric discharge, such as photoionization and/or thermal ionization processes. Deionization by recombination and/or diffusion can also be observed [47]. In addition to a high field stress, an initiatory-free electron rate can affect discharge events by a random time lag which depends on the amount of pre-ionization or irradiation of the gap [50].

2.3.5 Partial Discharges Mechanism

Three major types of discharges can be distinguished [51]. The first is internal discharge, which refers to discharge within dielectric insulation caused by gaseous inclusion or gas bubbles in liquids. The second major type of discharge is external discharge. Often known as corona discharge, it takes place in ambient air. The third type occurs along solid dielectric surfaces in ambient air. These discharges may bridge in a long gap distance and can erode solid insulation surfaces due to the high temperature in the discharge site [51, 52]. Their physical mechanism is related to the ionization processes induced by electron-avalanches as observed in [49, 53].

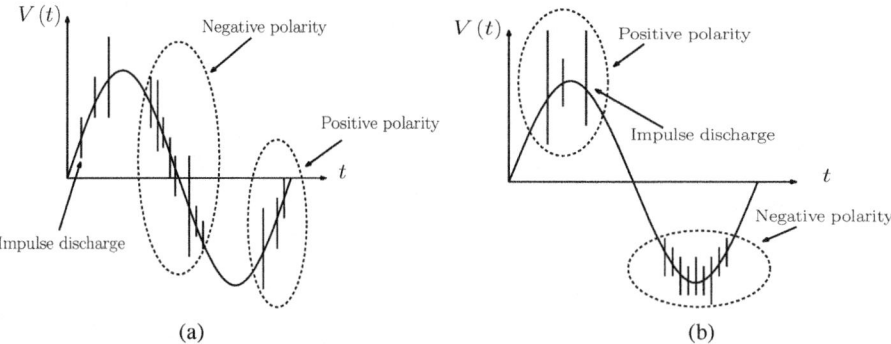

Fig. 2.5 Typical behaviour of impulsive noise induced by a discharge source. (**a**) Discharge along insulation surfaces. (**b**) Discharge in an air gap

Partial discharge (PD) phenomena are represented as stochastic processes in which amplitude, inter-arrival time between two successive impulses, and time occurrence are random variables. This can be explained by the variability of physical mechanisms such as the presence of ionizing radiation, fluctuations in gas density or gas decomposition in cavities or dielectric surfaces [54, 55]. Under AC voltages, PDs occur on every half-cycle of the applied voltage. Typical behaviours of impulsive noise from PD activity are presented in Fig. 2.5. PD events are superimposed upon the AC voltage.

When PDs take place along insulation surfaces, an asymmetric behaviour between polarities can be observed [52, 56]. Their amplitude and occurrence are higher at negative polarity than at positive as depicted in Fig. 2.5a. This behaviour can be found during slot discharge, surface tracking or gap-type discharge events [51, 52, 56]. For PDs in an air gap, the process occurs at the peak region of the applied AC voltage. At the negative polarity, impulses, known as Trichel impulses, can be observed with high repetition rates (between 50 and 100 kHz) and low amplitude, while impulses at the positive polarity, known as a pre-breakdown streamer, have low repetition rates with high amplitude as illustrated in Fig. 2.5b. This behaviour can be found in overhead power-lines [18, 19, 30, 31, 57].

These discharges can cause degradation and possible mechanical failure of electrical insulations [35, 47, 51]. They are also interference sources for TV and FM radio [30, 58]. Over the last four decades, PD detection and characterization methods have become important in the study of aging mechanisms and life-time analysis of HV equipment.

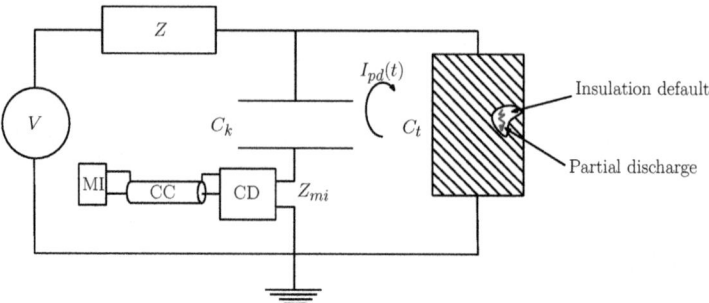

Fig. 2.6 Test circuit for partial discharge detection (*Courtesy of IEC-60270* [59])

2.3.6 *Measurements and Characterization of Partial Discharge Sources*

2.3.6.1 Measurement Techniques

Partial discharges can generate electrical pulse currents, dielectric losses, electromagnetic radiations, chemical reactions, and more. In this chapter, we present methods for PD detection and measurement based on impulse currents and electromagnetic radiation. These two methods are commonly used [47].

2.3.6.2 PD Currents Impulses

PD impulse currents can be measured by connecting the test object to a high-voltage source. The circuit for a PD test includes an impedance Z and a coupling capacitor C_k in parallel to the test object as depicted in Fig. 2.6. The test object can be seen as a capacitor C_t. During the short period of the partial discharge, C_k is a storage capacitor which can release a PD current impulse $I_{pd}(t)$ between C_k and the test object C_t.

The PD pulse apparent charge q can is given by:

$$q = \int I_{pd}(t)dt \tag{2.2}$$

when $C_k \gg C_t$. It is said apparent because the measured charge q is not equal to the charge locally involved at the discharge site [47, 59]. Thus, the true charge cannot be measured directly. However, a calibration procedure can be used to improve the measurement of PD quantities.

Most PD measurement systems based on PD impulse currents are integrated into the test circuit as shown in Fig. 2.6. The circuit includes a coupling device with its impedance Z_{mi}, and the measurement instrument is linked by a connecting cable. The coupling device can be a passive filter, such as a parallel RLC resonance circuit

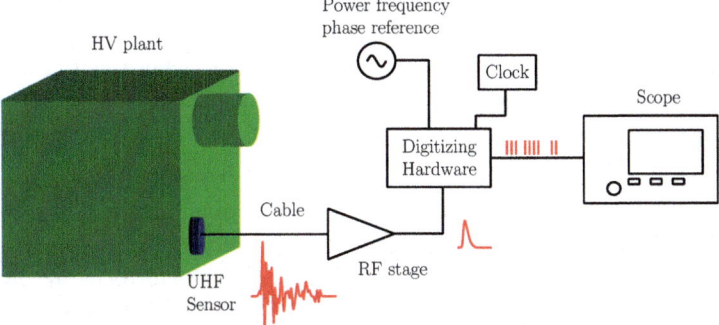

Fig. 2.7 Typical measurement setup using UHF PD detection (*Courtesy of Judd et al.* [41])

as was used by [35]. The input current and the output voltage are linked by the impedance of the filter when the apparent charge of PD pulse can be measured. Low- and high-frequency currents can be filtered by adjusting the parameters of the RLC circuit.

2.3.6.3   PD Electromagnetic Radiations

EM radiations from PD activity can be detected by passive devices such as antennas or sensors. They can be detected in the ultra high frequency (UHF) range [21, 22, 41, 60–62]. There are several advantages to these methods: sensors do not need electrical connection to the high voltage circuit, they ensure a better signal-to-noise ratio, and their use allows failure location to be determined by using PD localization methods [61]. UHF PD measuring methods are mostly used in gas-insulated substations (GIS) in which HV equipment is contained in a sealed environment with sulfur hexafluoride gas (SF6). A typical measurement setup is presented in Fig. 2.7. Signals from the UHF sensor are filtered, amplified by RF stages and digitized by digitizing hardware to extract PD quantities.

Electromagnetic radiations from PD can be measured using antennas. A typical measurement setup includes a wideband antenna, RF filters and amplifiers. Impulses are recorded by a scope over a very large frequency range. Measurement campaigns conducted in [18, 19] have used a wideband antenna to measure corona and gap-type discharges in a 60 Hz to 10 GHz frequency range on 2.4–765 kV overhead power-lines under AC voltages. It has been found that the power spectral density (PSD) of these interferences have a form of $1/f^{\gamma}$ where γ is the exponent characterizing the decay over frequency [18, 19, 21]. The waveform is impulsive with transient effects [21, 63, 64].

These discharges are also EMI sources for RF communications [58, 65–67]. PD measurement and detection methods using antennas are appropriate for RF channel characterization in substation environments. In the literature, the physical mechanism of PD and its electromagnetic radiation is not explicit and the characterization is often incomplete.

2.3.6.4 Characterization of PD Impulses

PD phenomena are a category of stochastic processes whose characteristics can be described as time-dependent random variables [55]. PD impulses are then characterized by:

- The repetition rate, which is the total number of PD impulses occurring within an arbitrary time interval. With AC voltages, this time interval can be a cycle of the applied voltage. Thus, the repetition rate is defined as the number of PD impulses per cycle. The repetition rate can be also defined as the number of impulses during the positive or the negative polarity of the AC voltage;
- The inter-arrival time (IAT), which is the time between two consecutive PD impulses. Under AC voltages, IAT can be measured at the positive or at the negative polarity;
- The inter-impulse time (IIT) which is the time between the end of an impulse and the beginning of the next impulse;
- The occurrence which is the time in which a PD occurs under the operating voltage. Under AC voltages, the PD process is cyclostationary as the PD occurs at every half-cycle of the applied voltage;
- The duration of an impulse. PD can have different physical processes in which the duration can be variable;
- The amplitude of a discharge, which can be measured by the apparent charge of the PD phenomena as recommended by the standard IEC 60270 in [59]. In this case, impulse currents measurement methods are used. For electromagnetic radiations, the average energy can be calculated.

Under AC voltages, phase-resolved partial discharge (PRPD) representation is commonly used in PD characterization [52, 55, 56]. This is a three-dimensional statistical representation of a PD process in which probability distributions of PD events and amplitude are plotted on the phase of the operating voltage. One can identify typical patterns of PDs with PRPD as depicted in Fig. 2.8. PD measured on a stator bar in which PDs occur at every half cycle of the applied voltage. The PD impulse currents measurement method is used in this figure.

2.3.7 Partial Discharge Modeling

PD modeling is an extensive research area in which physical and statistical models have been investigated [68–72]. PD models can be classified as either physical PD models or statistical PD models.

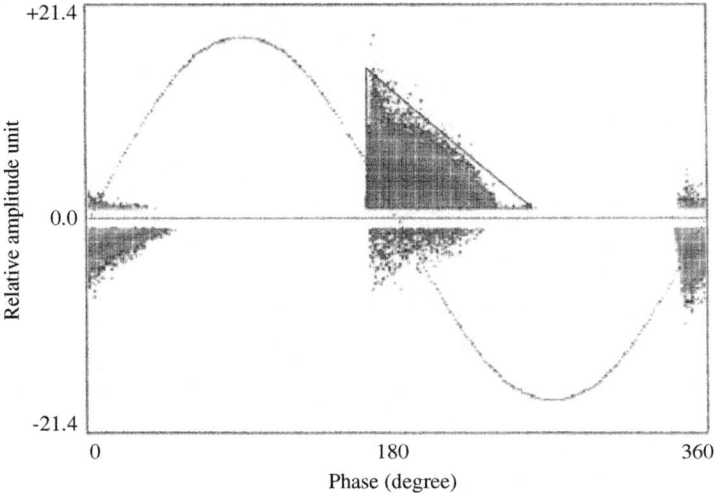

Fig. 2.8 PRPD pattern of PD measured on a stator bar (*Courtesy of Hudon and Belec* [52])

2.3.7.1 Physical PD Models

Physical PD models have been developed to extend knowledge about the evolution
of PD activity. Researchers have shown that these models use the physical approach
in which PD activities are simulated in spherical voids [68, 69], or in stator
bars [70]. Essentially, the electric field in the PD site is calculated by numerical
methods in each model. Based on physical criteria such as temperature, pressure
and the radius of the cavity, the critical value for discharge is determined. When
the electric field in a given PD site is higher than the critical value, a discharge
occurs in which PD charge amplitude can be calculated physically. These models
can reproduce the cyclostationary process under AC voltages. However, the PD
charge amplitude cannot provide any information about the induced electromagnetic
radiations. Waveforms and spectra of RF signals from the PD are not taken into
consideration.

2.3.7.2 Statistical PD Models for Wireless Channels

Middleton Class A and α-stable noise models are commonly used to reproduce
EMIs from PD in wireless communication channels [71, 72]. Based on measurement
campaigns in substations or in laboratories, statistical parameters are estimated from
data using statistical methods [73–77]. The resulting probability distributions are
compared to the experimental results. The Kullback-Leibler (KL) divergence is
used to quantify the difference between two probability density [78]. This is a non-
symmetric measure of the distance between two arbitrary probability densities f_1
and f_2. For discrete probability densities, the KL divergence of f_2 from f_1 is defined

by:

$$D_{KL} = \sum_i f_1(i) \ln \frac{f_1(i)}{f_2(i)} \tag{2.3}$$

If f_1 and f_2 are identical, then $D_{KL} = 0$. In addition, a Kolmogorov-Smirnov (KS) test for the null hypothesis that two sample data are from the same distribution. The test statistic is:

$$D_{KS} = \sup_x |F_1(x) - F_2(x)| \tag{2.4}$$

where $F_1(x)$ and $F_2(x)$ are the empirical cumulative distribution functions (CDF). $\sup_x$ is the supremum. If two sample data are from the same distribution, then D_{KS} converges to zero [79].

An example of a typical PD impulse measured in a substation is depicted in Fig. 2.9. Gaussian, Middleton Class A and α-stable noise models are used to reproduce data. Parameters are estimated from the data using statistical methods [73–77]. PDFs and resulting noise samples are plotted for comparison. It can be seen that impulsive noise produces a heavy-tailed distribution. As opposed to the Gaussian distribution model, the heavy-tailed behaviour is taken into account in Middleton Class A and α-stable noise models. Therefore, the KL divergence values are smaller than those in the fitted Gaussian noise model as seen in Table 2.1.

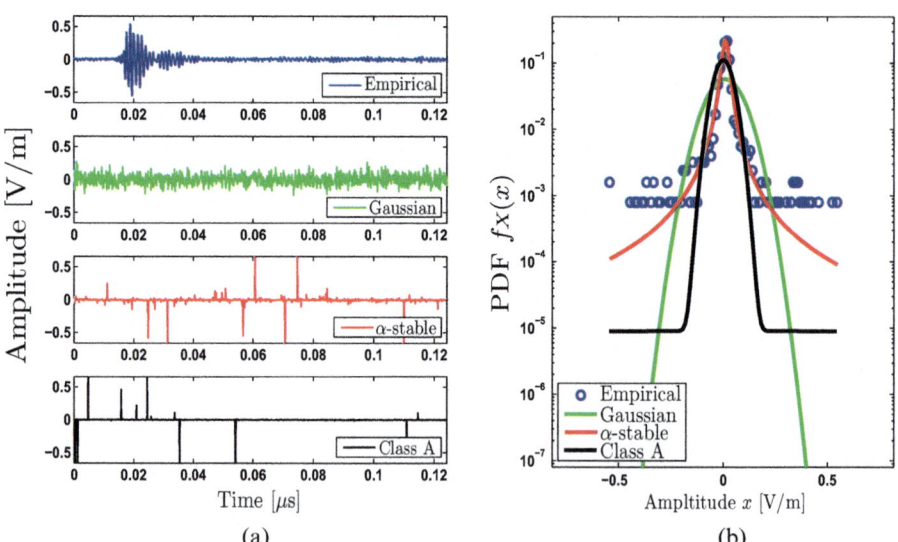

Fig. 2.9 Impulsive noise with fitted statistical models. (**a**) An impulsive noise waveform. (**b**) PDF of amplitude

Table 2.1 The goodness-of-fit of impulsive noise models

Test statistics	Gaussian	Middleton class A	α-Stable
D_{KL}	1.18	0.596	0.08
D_{KS}	0.56	0.24	0.195

Unfortunately, using memoryless models can be limited because impulsive samples are generated independently. By conducting the KS test for a significance level of 5%, p-values can be derived from the test statistics values that are tabulated by Massey, Jr. in [79]. These p-values are lower than 0.05 for all of these noise samples impulsive noise models. As a result, the test indicates the rejection of the null hypothesis that noise samples from these statistical impulsive noise models and the measured impulsive noise cannot come from the same distributions at a 95% confidence interval. Moreover, these statistical models exhibit lack-of-fit because the resulting waveforms generate *iid* noise samples which are not observed in the data. Indeed, we can see that the PD impulse occurs in bursts with transient effects and damped oscillation. Using KL divergence to validate statistical models from measurements is not accurate because the spectrum characteristics of EMIs are not taken into account.

2.4   Characterization and Impulsive Noise Models

In the literature, models for impulsive noise are widely developed and used to extend the existing knowledge regarding the nature of impulsive noise sources and communication performance analysis. In this section, characterization methods and existing impulsive models are reviewed.

2.4.1   A Statistical Characterization of Impulsive Noise

In practice, measured impulsive noise consists of impulsive events and additive background noise. The latter is produced by thermal noise from the measurement setup as well as ambient noise, which is generated by many interferences below the level of impulses as depicted in Fig. 2.10. Since these impulsive waveforms are characterized by their short duration and high amplitude, they can be detected using a simple threshold. A more sophisticated technique can be used to extract impulses from overall background noise to yield an estimation of power spectral density. This will be detailed in Chap. 3.

Impulsive interferences can be characterized in one of two ways. Using first-order statistics, statistical distribution can be calculated using duration, the inter-arrival time between two consecutive impulses, amplitude, and energy. Assuming that impulsive noise is a non-stationary random process, short-time analysis is used

Fig. 2.10 Typical example of impulsive noise with background Gaussian noise

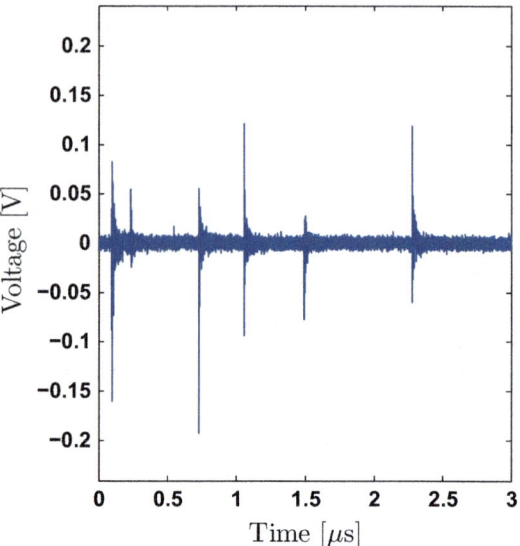

in the analysis of first-order statistics. Second-order statistics utilizes power spectral densities, and a Spectrogram, a Wigner-Ville distribution, or a scalogram can be used to estimate PSD [80–82].

2.4.2   Impulsive Noise Models

Impulsive noise can be modeled as a succession of short impulse waveforms with background noise over a large observation time. Various approaches can be used to represent impulsive waveforms. In [24], the author classifies these interferences into three categories as follows:

- EMIs can produce negligible transient waveforms in a typical receiver. It is denoted by Class A noise. In terms of receiver bandwidth Δf_R and duration of interference sources T_U, the Class A noise model assume that:

$$T_U \, \Delta f_R \gg 1 \qquad (2.5)$$

- when EMIs are characterized by transient effects, they are denoted by Class B noise when the receiver bandwidth and the duration of interference sources are:

$$T_U \, \Delta f_R \ll 1 \qquad (2.6)$$

- class C is a mixture of Class A and B in which Class B is predominant compared to Class A.

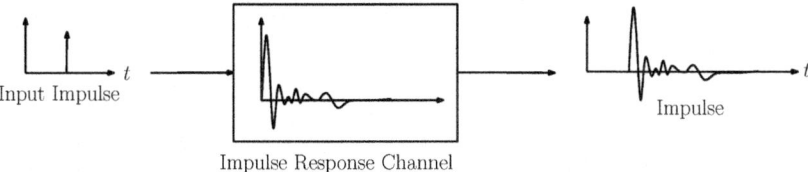

Fig. 2.11 Impulsive noise modeled by a filter

An impulse can be modeled as a unit-impulse function $\delta(t)$ with an infinitesimal time width or duration. However, this does not exist in practice because impulsive interferences have a non-zero finite duration in real-life.

A channel impulse response can be used to model various impulsive noise waveforms such as damped oscillation with transient effects, etc. As presented in Fig. 2.11, a typical impulsive waveform can be modeled by using a filter in which the input model is an ideal impulse. In this condition, a power spectral density can be determined.

For a given impulse time-occurrence t_n, the resulting waveform can be written as a convolution product between the input impulse and the filter, such that:

$$u(t) = h_m(t) * \delta(t - t_n) \tag{2.7}$$

where the impulse response of the filter is given by $h_m(t)$ and $*$ is the convolution product operator. Physically, the impulse response includes the propagation channel, the impulse response of both the emitted source and the measurement setup. The power spectral density of the resulting impulse is only determined by the power spectral density of this filter.

2.4.3   Probability Models of Impulsive Noise

Impulsive noise can be characterized and modeled by its instantaneous amplitude probability density function. This representation is commonly used in the study of communication performance. Over a long observation time, the resulting noise is a superposition of many independent sources with a random number of impulsive radiations and their locations are randomly distributed in space. The overall noise is given by:

$$x(t) = \sum_k u_k(\theta_k, t) \tag{2.8}$$

where $u_k(\theta_k, t)$ is the kth interferer, and θ is a set of random variables characterizing the interference.

In the literature, many existing impulsive noise models have been developed [23, 24, 83, 84]. They can be classified as either memoryless noise or noise models with memory [84].

2.4.3.1 Memoryless Models

Bernoulli-Gaussian Model

The Bernoulli-Gaussian model is the simplest way to represent impulsive noise since it is a mixture of two zero-mean Gaussian distributions weighted by a single Bernoulli probability λ [84, 85]. The probability density function (PDF) of the samples $f(z|\sigma_0^2, \sigma_1^2, \lambda)$ for this model is:

$$f(z|\sigma_0^2, \sigma_1^2, \lambda) = \frac{\lambda}{\sqrt{2\pi\sigma_0^2}} \exp(-\frac{z^2}{2\sigma_0^2}) + \frac{1-\lambda}{\sqrt{2\pi\sigma_1^2}} \exp(-\frac{z^2}{2\sigma_1^2}) \qquad (2.9)$$

In general, one Gaussian distribution with variance σ_0^2 characterizes the background noise and the other one, the impulses with variance σ_1^2. The concept of this model is simple because it assumes there is only one source of impulsive noise that generates i.i.d. impulses for a one-sample duration. According to (2.9), the model needs only three parameters, which are the background noise variance σ_0^2, the impulses variance σ_1^2, and the Bernoulli probability λ (the probability to be in an impulsive state is $1-\lambda$). The model can also be represented by a Markov chain (Fig. 2.12 with state 0 the background state and state 1 the impulse state), with the following probability matrix:

$$T = \begin{pmatrix} \lambda & (1-\lambda) \\ \lambda & (1-\lambda) \end{pmatrix}$$

The Bernoulli-Gaussian model is commonly used in simulations to represent impulsive noise when very little information is available [86–89]. The only information required by the model are the rate of impulses (impulses per second) and the variances of the impulse and the background noise.

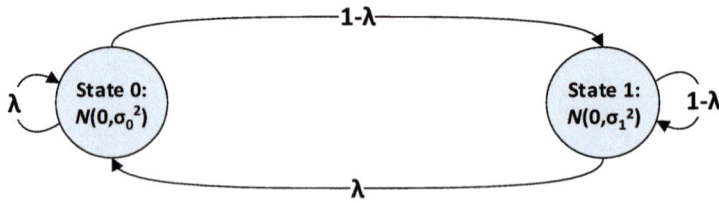

Fig. 2.12 Bernoulli-Gaussian model represented in a Markov chain model, where $\mathcal{N}(0, \sigma_i^2)$ is the zero-mean Gaussian distribution with variance σ_i^2 of the state i, $i = 0, 1$

Even though this model does not generate impulsive noise as observed in substations, it can represent the impulse occurrence with their amplitude maxima and provide the lowest complexity in terms of parameter estimation. For future applications, such as simulating noise samples or embedding the model into a receiver, the Bernoulli-Gaussian model is also one of the least complex.

Poisson-Gaussian Noise

We assume the overall noise to be a superposition of independent EMI emissions modeled as a Poisson process for a large observation time T, while the amplitude distribution of impulses is a Gaussian noise process. From Eq. (2.8), the resulting noise is given by:

$$x(t) = \sum_{k=1}^{N_I(t)} a_k \delta(t - t_k) + n(t) \tag{2.10}$$

where a_k is the random amplitude of the kth impulse and the impulse occurrence time is given by t_k. $N_I(t)$ is the number of impulses driven by a homogeneous Poisson process. The probability occurrence of the kth impulse in the observation time T is given by:

$$f(k; \lambda; T) = \Pr\left[N_I(t + T) - N_I(t) = k)\right] = \frac{e^{-\lambda T}(\lambda T)^k}{k!} \tag{2.11}$$

where λT is the average number of emissions occurring within the observation time and λ is the density of emissions. The probability density function (PDF) of impulsive noise is given in [84] in a small time interval Δt by:

$$f_I(u; \Delta t) = (1 - \lambda \Delta t)\,\delta(u) + \lambda \Delta t \frac{1}{\sqrt{2\pi}\,\sigma_u} e^{-u^2/2\sigma_u^2} \tag{2.12}$$

where σ_u is the variance of the impulsive component. Since the background noise and impulsive noise are a sum of independent random processes, the resulting PDF is written as the convolution product of the background noise PDF and the impulsive noise PDF such that:

$$f_{P+G}(x) = f_I(x) * f_n(x) \tag{2.13}$$

where $f_n(x)$ is the Gaussian distribution of the background noise; and

$$f_n(x) = \frac{1}{\sqrt{2\pi}\,\sigma_n} e^{-x^2/2\sigma_n^2} \tag{2.14}$$

where σ_n is the variance of the background noise component.

Middleton Class A and B Noise

Extended to Poisson-Gaussian models, physical-statistical impulsive noise models are proposed by [24, 25, 76]. Based on the Poisson field of interferers, the resulting noise is a superposition of independent EMI sources randomly distributed in space which results in emissions that can be activated within a given observation time T. The PDF of overall noise can be derived for Class A noise as:

$$f_{P+G}(x) = e^{-A} \sum_{m=0}^{\infty} \frac{A^m}{m! \sqrt{2\pi \sigma_m^2}} e^{-x^2/2\sigma_m^2} \qquad (2.15)$$

where $\sigma_m^2 = \frac{m/A + \Gamma}{1+\Gamma}$.

- the parameter A is the impulse index. This is the average number of emissions for a given observation time.
- the parameter Γ is the ratio between the level of the background noise and the level of the impulsive noise component where $\Gamma \in \left[10^{-6}, 1\right]$.

In literature [90, 91], A and Γ are typically in the range of $\left[10^{-2}, 1\right]$ and $\left[10^{-6}, 1\right]$ respectively.

The Class A PDF can be seen as a weighted sum of Gaussian probability densities. This model is widely used due to its canonical form, and accordingly, this is an analytically tractable model for narrowband interferences in which the transients in the typical receiver can be neglected.

Class B interferences with transient effects is a broadband noise model which is thus more complex than the Class A noise model [24]. The interference noise model is physically coherent, due the presence of the Gaussian noise component (background noise). However, six parameters and an empirical parameter have to be considered in this model, which is more difficult to use.

α-Stable Noise

As an alternative to the Class B interference model, α-stable noise model is used to approximate the Middleton Class B model [92]. The α-stable distribution was introduced by [93] in his study of normalized sums of *iid* random variables. By definition, a random variable X is said to be stable if for any positive constant numbers a and b, there is a random variable $aX_1 + bX_2$ has the same distribution as $cX + d$ for a positive number c and a real number d. Note that X_1 and X_2 are two independent copy of the random variable X.

By letting X be a stable random variable, the noise model is defined by its characteristic function (c.f.) $\mathscr{M}_X(j\xi)$ given by:

$$\mathscr{M}_X(j\xi) = \mathbb{E}\left[e^{j\xi X}\right] = \begin{cases} \exp\left\{j\xi\mu - |\sigma\xi|^\alpha \left(1 - j\beta \, \mathrm{sign}(\xi)\tan(\pi\alpha/2)\right)\right\}, & \alpha \neq 1. \\ \exp\left\{j\xi\mu - |\sigma\xi|^\alpha \left(1 - j\beta \, \mathrm{sign}(\xi)\frac{2}{\pi}\log|\xi|\right)\right\}, & \alpha = 1. \end{cases}$$
$$(2.16)$$

where $\mathbb{E}[\cdot]$ denotes the expectation, $0 < \alpha < 2$ is the stability index, μ is a location parameter real value, $\sigma \geq 0$ is a scale factor and β is the skewness parameter where $-1 \leq \beta \leq 1$. The model can take into consideration the heavy-tailed behaviour induced by impulsive events in the distribution. By using the inverse Fourier transform, the PDF is obtained. However, closed forms are difficult to derive. Only a few stable distributions can be written in closed forms for specific parameters of α for:

- $\alpha = 2$, the PDF is the Gaussian distribution, $X \sim S_2(\sigma, \beta = 0, \mu)$, where the mean is given by μ and its variance is 2σ. Note that the scale factor σ is not the same as the variance of the Gaussian noise;
- $\alpha = 1$, the PDF is the Cauchy distribution, $X \sim S_1(\sigma, \beta = 0, \mu)$ and
- $\alpha = 1/2$, the PDF is the Levy distribution, $X \sim S_{1/2}(\sigma, \beta = 1, \mu)$.

A study in [24] has reported that the α-stable model is physically incomplete since the background noise modeled as a Gaussian noise is not considered. These memoryless models have the advantage of being analytically tractable when parameters of these PDFs can be estimated by using statistical methods [73–77]. Based on these impulsive noise models, performance analysis of communication systems have been investigated [94–97].

2.4.3.2 Impulsive Noise with Memory: Burst Noise

Markov-Middleton Model

Memoryless impulsive noise models generate *iid* noise samples, in which an impulse sample occurs at random instants of time. However, in practice, most environments, such as high-voltage substations or power-line distribution systems, experience impulsive noise in bursts [21, 23, 98]. Markov chains have been investigated for impulsive noise modeling in instances when this bursty behaviour of impulsive noise can be reproduced.

Recently, a Markov-Middleton model has been proposed in [83]. This model uses the Middleton Class A noise model but includes an extra parameter to ensure the presence of impulse memory with a hidden Markov model, as shown in Fig. 2.13. Each state generates a noise sample with specific PDF. A transition state is considered when duration within the state is null. The correlation between noise samples is ensured by the probability p_x, which is independent of Middleton Class A parameters. The entering state $m = \{0, 1, 2, 3, \cdots\}$ from the transition state is controlled by the probability p_m such that:

$$p_m = \frac{p_m'}{\sum_{m=0} p_m'} \tag{2.17}$$

Fig. 2.13 Markov-Middleton impulsive noise model with three states of Markov chain (*Courtesy of Ndo et al.* [83])

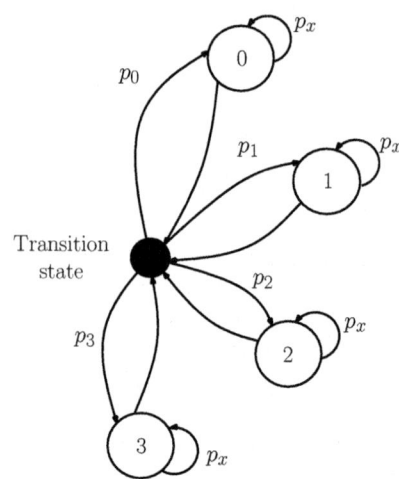

where $p'_m = e^{-A} A^m / m!$, according to the Middleton Class A PDF model. Each state generates noise samples from a zero-mean Gaussian noise where its variance is given by $\sigma_m^2 = \frac{m/A + \Gamma}{1+\Gamma}$.

The model is a truncated version of the Middleton Class A PDF, which is analytically tractable, and the time correlation of bursty impulsive noise is taken into consideration. However, the number of states in Markov chains might be difficult to define, and the resulting impulsive noise samples are still uncorrelated. Therefore, transient and damped oscillation effects are neglected.

Partitioned Markov Chain Model

Alternatively, partitioned Markov chain (PMC) models can be used to reproduce memory effects of impulsive noise. A PMC model has been developed for broadband power-line communications in [23]. This model can reproduce the bursty behaviour of impulsive noise in PLC. Simulation results show that the model fits measurements. An example of PMC for impulsive noise is depicted in Fig. 2.14. The K states are partitioned into two groups; A denotes the absence of an impulsive component, and B, the presence of an impulse event. This is the generalization of the Gilbert-Elliot model for bursty impulsive noise modeling [99, 100]. The latter considers only two states of the Markov chain. The two groups can be described independently by transition probability matrices for impulse free states and for impulse states. In addition, transition states are introduced in order to regulate the transition from impulse free states to impulse states and vice versa.

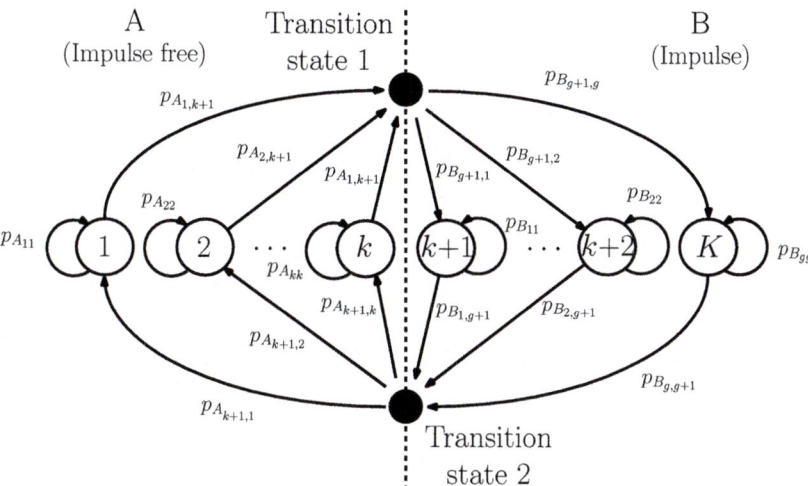

Fig. 2.14 Partitioned Markov chain for asynchronous impulsive noise (*Courtesy of Zimmermann and Dostert* [23])

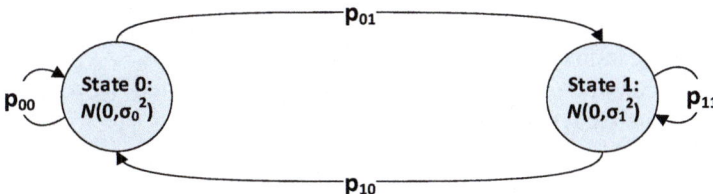

Fig. 2.15 Bernoulli-Gaussian model with memory

Two-State Markov Chain

Memory can be implemented in the Bernoulli-Gaussian model by selecting different transition probabilities (Fig. 2.15) to increase the chance to generate two or more consecutive impulses. This model is a particular case of the partitioned Markov chain [23, 85] since one state represents the background noise and the other represents the impulses. We call this model either two-state Markov chain (MC2) or Bernoulli-Gaussian with memory (BGM).

MC2 can generate impulses with samples that follow a Gaussian distribution with a variance larger than the background noise variance [101]. The duration of the impulse is implemented by the probability to remain in the impulse state (state 1 in Fig. 2.15). As for the partitioned Markov chain of Zimmermann, the impulse generation of MC2 is implemented by the transition probability p_{01} from the background state (state 0) to the impulse state (state 1). The time-correlated nature of impulsive noise that shapes the impulses in damped oscillations is not modeled; as a consequence, the power spectrum of the impulses is flat, due to the use of a single Gaussian distribution to represent the impulse samples.

These statistical impulsive noise models can reproduce EMIs in various impulsive environments. However, they do not take into account the specificity of the electromagnetic environment in substations; In other words, they cannot provide any information linking the physical characteristics of HV installations to the induced radio interference spectrum. The next section focuses on the electromagnetic interferences in substations.

2.5 Wireless Communications in Substations

2.5.1 Communication Channels in Presence of Impulsive Noise

For several years, there has been a growing interest in both broadband power-line communications and wireless communications within industrial environments. Unfortunately, various EMIs especially impulsive noise can interfere with communication systems. Many papers show that the receivers perform poorly [67, 72, 95, 102–105]. Due to extensive research, more robust optimum and suboptimum receivers that function even in the presence of impulsive noise have been developed, along with mitigation techniques [83, 94, 95, 106, 107].

Typically, communication channels include a transmitted signal and additive interferences such that the received signal is written as:

$$x(t) = s(t) + n(t) + \sum_k u_k(\theta_k, t) \qquad (2.18)$$

where $s(t)$ is the transmitted signal, $n(t)$ is the receiver's thermal noise and $u_k(\theta_k, t)$ is the kth interferer as stated in Eq. 2.8. If we assume that $x(t)$ is a sequence of independent and identically distributed (*iid*) random variables with the same expected value and a finite variance, then, the sum of overall received signals converges in distribution to a Gaussian distribution. From this assumption, classical coding theory is well-known [108, 109]. However, when the receiver is corrupted by few number of interferences highly impulsive, the probability density is heavy tailed. Therefore, the assumption of Gaussian distribution no longer applies and the classical decoding scheme and detection rules for additive white Gaussian noise (AWGN) channel are more likely to cause erroneous decisions on the transmitted message.

2.5.2 Wireless Technologies

Wireless communication is a way to exchange data over different ranges with advantages such as easy deployment and mobility. Wireless technologies can be used for multiple purposes; we can find wireless standards that differ in terms of bandwidths, carrier frequencies, transmission coverage, transmission power and receptor sensitivity. We classify, in the following list, the main wireless technologies that can provide the communication systems for substation IED networks.

1. **Bluetooth** (IEEE 802.15.1) is designed for short range (10–20 m) with low power consumption and has a maximum throughput on the order of 1.5 Mb/s. This technology uses the 2.4–2.485 GHz band and can be used for monitoring applications, communications with computers, IEDs, hands-free kits, etc.
2. **ZigBee** (IEEE 802.15.4) is designed for medium range (10–100 m) for the following bands: 2.4 GHz in most jurisdictions worldwide; 784 MHz in China, 868 MHz in Europe and 915 MHz in the USA and Australia. This technology offers 16 channels spaced by 5 MHz with 2 MHz bandwidth.
3. **Wi-Fi** (IEEE 802.11 a,b,g,n,ac) is designed for medium/long ranges (100–200 m on average). This technology is available for 2.4 and 5 GHz carrier frequency, with bandwidths of 20 or 40 MHz depending on the version of the standard.
4. **Wimax** (IEEE 802.16) is designed for Broadband Wireless Access (BWA) over very long range (50 km) and with a very large throughput (100 Mb/s). This technology uses 2.3, 2.5, 3.3, 3.5 and 5.8 GHz carrier frequencies depending on the county.
5. **LTE** has emerged since 2008 as a new long range (50 km) wireless mobile communication standard. This technology offers large bandwidth from 20 to 100 MHz for carrier frequencies at 700 MHz and 1.8 GHz in North America [11].

None of the wireless technologies in the list above are designed for substation environments, nevertheless some industrial projects have managed to install wireless sensors in the vicinity of equipment similar to those found in substations [110]. We will present the possible applications of such technologies in industrial projects and we will explain the conditions to ensure reliable wireless communications.

2.5.3 Existing Systems for Wireless Communications in High Voltage Environment

Communications in high voltage environments are important for applications such as wireless sensors networks in ship engine room, Bluetooth sensors for current transformers or wireless data transfer from within substations to the outside world [110]. The specific applications of wireless communication require selecting a wireless technology that is able to transmit data with a specific rate and ensuring that the environment does not affect the communication performance too negatively.

Wireless technologies are not designed for high voltage environments, so the network might require installing the sensors in the less noisy locations. The adaptation of wireless sensors to any environment must consider the factors that might disturb the transmission, such as congestion, electromagnetic interferences (EMI), spectrum availability.

Hence, one should envisage adapting commercial wireless technologies to high voltage environments in order to communicate without constraint. This section presents some applications of commercial wireless technologies in power environments and highlights the necessity to find a tradeoff between the transmission reliability and the sensors location.

Wireless technologies have long been used by power management companies for operations, such as microwave systems, Multiple Address Systems radios (MAS), spread spectrum radios, limited satellite applications (VSAT, paging and GPS), limited mobile radio systems. However, these applications are rarely used inside substations. The use of wireless communications in power environment is possible but surely limited in applications; wireless technology enhances reliability through the use of information diversity, which means that the data can be lost for a period of time and the system can manage to recover them (information recovering) as soon as the transmission is working . However, the delay induced by the information recovering in power environments limits the use of wireless technologies to non-critical operations. We describe here some examples of those applications in environments similar to substations that are gathered in [110].

Current transformer sensors connected through Bluetooth constitute a wireless application to control current transformers based on data coming from power lines [110]. Accurate current sensors can easily be deployed on power lines and give information about them. The sensors are connected via Bluetooth to a station located in a building, which is also connected with cables to the active current transformer using a three-phases meter. Hence, the station controls and exchanges data with the equipment.

Crossbow Wireless Network is used in *British Petroleum Company (BP)* projects to facilitate maintenance of their ship machines by monitoring the wear index based on vibration data [110]. Depending on the data, BP is informed if a maintenance operation is required or not. The sensors are installed on the machines and connected to a signal processor that converts the results to digital data. Thereafter, the data are transmitted through the Crossbow wireless network. The metallic structure of the ship decreases the transmission performance, so the wireless network is meshed to ensure communication between each radio node. Data are transmitted node to node on a obstacle-free route until they reach the main gateway on the ship main network.

Laptops used to connect to equipment in substation: Schweizer Engineering Laboratories (SEL) have created a communication module that, once connected to a PC serial port, is able to communicate over wireless links (802.11b) with a controller in a substation. The main purpose is to allow the substation staff to interact with the equipment from the outside [110].

Many projects study the performance of wireless networks in substations working under 20–500 kV voltage for monitoring, protection and control [7, 14–17]. The deployment of wireless IED confirms that existing wireless technologies are not ready for a direct application [14, 16, 17] and designing industrial wireless sensors would allow reliable transmissions without limiting the network to "safe areas" in terms of noise. Battery saving is another issue for wireless sensors and industrial wireless technologies, inter alia, would also allow the network to remain robust without increasing the transmission power [7, 16].

In all wireless applications in high voltage environments, a tradeoff is made between the network location, the number of nodes, the delay and the throughput for a specific application to ensure transmission [111–114]. Interference and reflections from metal structures are sources of degradations in communication quality that requires the use of more sensors and their placement in areas less exposed to electromagnetic noise. The projects described in this section use short communication distances and the voltage used by surrounding equipment does not exceed 20 kV, which is below than 230–735 kV voltage range of typical substations; then the problem is not solved but avoided. Commercial wireless communications are not adapted to offer a reliable transmission in high voltage environment, therefore new modifications should be added to wireless devices. The main objective of these techniques is to make sure the sensors can communicate reliably wherever they are installed.

2.6 Summary

Substation equipment generates impulsive noise and existing wireless technologies are not designed to ensure robust communications. Although some adaptations are made in less harsh environments, such as ship machinery or distribution substations, the transportation substations generate impulsive noise that threatens the reliability of a wireless communication in a scenario of IED sensors network. Many models exist to represent impulsive noise, and according to noise observations, it is important to model the correlation between the samples. A reliable model must consider the time-correlation of the noise in order to represent accurately the environment of substations. In a wide-band framework, the assumptions made by other researchers do not hold anymore; i.i.d. noise samples or the use of a specific wireless technology cannot be assumed. The correlation between the samples can only be replicated by a model with memory and an optimum receiver can take advantage of this correlation to calculate the appropriate metrics and mitigate the impact of the impulses. In order to reach such a goal, the next chapter reports measurement campaign that will illustrate the nature of the noise and help for the design of noise modeling.

Chapter 3
Impulsive Noise Measurements

The phenomenon of impulsive noise is generated by partial discharge (PD) sources in high-voltage substations. A PD generates a current impulse, acoustic noise, visible and ultraviolet (UV) light and electromagnetic radiation, and accordingly its presence can be detected via several measurement methods. In this chapter, PD measurement methods are based on the detection of electromagnetic radiations. These PD instrument detectors have the advantage to be non-invasive measurement methods for the high-voltage (HV) equipment as well as one can assess impulsive electromagnetic interference (EMI) threats to wireless communication systems.

This specific radio noise is a source of interference for the radio communication systems. In the literature, electromagnetic radiations from PD activity have been measured in several substations [18, 19, 21, 45, 63]. However, the measurement setup employed does not cover the frequency range used by conventional wireless communications, and so the impulsive noise characterization is often incomplete. Inspired by related works in PD current measurements and characterization [52, 54–56], a characterization process is developed in which the impulsive electromagnetic radiations from PD activity are fully characterized by the amplitude of the power density, inter-arrival time, occurrence and spectrum.

In substations, impulsive noise is composed of a Gaussian background noise and impulses that are shaped with a damped oscillating waveform [20, 21, 40]. If we consider a sampled impulsive noise, the waveform of the impulses implies that the impulse samples are time-correlated. Such information can be used by a receiver to mitigate the impact of the noise on wireless communications. In this chapter, we provide a non-invasive measurement and detection method for the characterization of the electromagnetic radiations induced by PDs. Impulsive noise measurements can provide all information required for modeling the noise induced by PD, which requires that the measured impulses contain enough samples to observe the noise pattern described.

© Springer International Publishing AG, part of Springer Nature 2019
B. L. Agba et al., *Wireless Communications for Power Substations:
RF Characterization and Modeling*, Wireless Networks,
https://doi.org/10.1007/978-3-319-91328-5_3

Measurement campaigns can provide different information that can be used to enhance wireless communications in substations. First, measuring impulsive noise in substations helps to identify the behavior of the impulses in terms of amplitude, duration, generation, waveform and spectrum. Secondly, the measured noise can be used for comparing different noise models, such as existing models and those we plan to design. Finally, simulations of communication receivers under impulsive noise is more relevant if the system is subject to impulsive noise that corresponds to a real substation environment.

3.1 Objectives of the Measurement Campaign

The measurement campaign and deploying wireless communication in substations are relevant since it would provide information regarding the substation equipment, the electromagnetic environment for wireless communications and PLC. Smart Grid oriented projects, are connected or dependent to this research work; for this reason we provide results that can be useful to other industrial research works for both power systems and telecommunications departments.

Since we want to characterize impulsive noise on a wide band, a specific measurement setup is designed in order to collect impulsive noise samples in substations. Transport substations owned by Hydro-Québec work under a very high voltage, usually between 25 and 735 kV. By conducting measurements in substation with different voltages, the possible correlation between the equipment voltage and the significance of impulsive noise in terms of sample value and impulse occurrence rate can be studied.

A setup is designed in order to provide enough samples to ensure a reliable statistical study of impulsive noise, while being robust to the harsh environment of substations. To reach these goals, it is important to bear in mind:

- The length of the recorded noise sequence,
- The frequency band,
- The electrical protection of the measurement equipment.

3.2 Measurement Setup

The measurement setup is designed to measure impulsive noise in time domain and for a frequency band that contains many of the carrier frequencies of existing wireless communications. The setup is not only designed to measure signals in the ISM band, but also to record signals on a time window that is long enough (51.2 ms) to contain the samples required for a reliable statistical study.

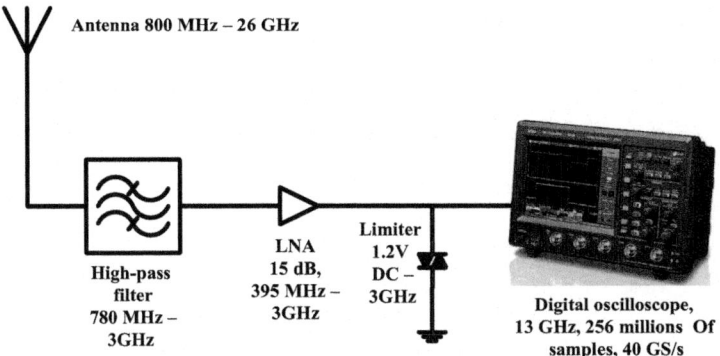

Fig. 3.1 Measurement setup of impulsive noise for ISM band: 780 MHz to 2.5 GHz

3.2.1 Design of the Setup

Measurements in time domain require the use of a digital oscilloscope connected to a wide band antenna. To measure impulsive noise in wide band, we propose the setup described in Fig. 3.1. The antenna and the oscilloscope are the two main components of the setup for they provide an electromagnetic-to-electric conversion of the noise and digital signal processing, respectively. The rest of the setup is composed of RF devices to ensure that measurements are performed in the frequency band of interest and that the signal coming from the antenna will not harm the input of the oscilloscope.

The list of setup components is shortly described below:

- An antenna that is unidirectional and polarized with a linear Gain on the band 780 MHz to 26.5 GHz;
- A 13-GHz digital oscilloscope with a maximum sampling rate of 40 Giga-Samples per second and a memory of 256 millions of samples;
- A high-pass filter, 780 MHz to 3 GHz. −100 dB attenuation and −0.5 dB of pass-band gain;
- A low noise amplifier (LNA) with a gain of 15 dB on the 395 MHz to 3 GHz band;
- A limiter that clips the signal to 1.1 V on the DC-3GHz band.

With the devices assembled as in Fig. 3.1, impulsive noise can be recorded on the 780 MHz to 2.5 GHz band with the oscilloscope configured as follows:

- Sampling frequency set at 5 Giga-Samples per second.
- Time window set at full memory (256 millions of samples), which corresponds to 51.2 ms.

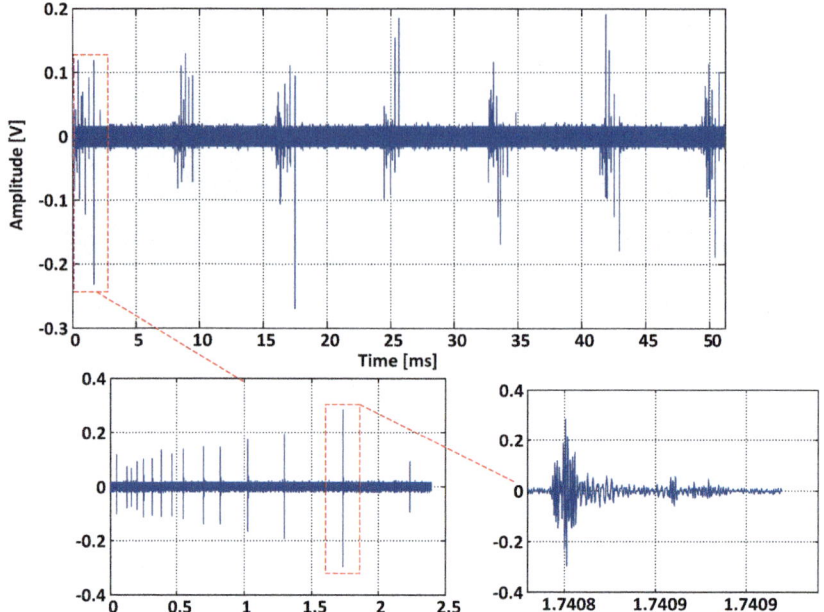

Fig. 3.2 Impulsive noise from Tesla-coil, 50 kV

3.2.2   Tests in Laboratory

In order to verify the reliability of the measurement setup, we perform an impulsive
noise measurement in a high voltage laboratory [115]. The setup is first tested using
a 50 kV Tesla-coil and a generator bar under high voltage.

Two scenarios are presented:

- A Tesla-coil is used in electro-technical laboratories to produce electrical
discharges in the air; it is usually used to ionize plasma. The Tesla-coil can be
used to generate Corona effect or external partial discharges when the tip is close
to a floating electrode, which creates an electrical arc from the electrode of the
coil to a floating electrode. The impulsive noise produced by the Tesla coil is
recorded and an example is presented in Fig. 3.2.
- A generator bar is used in controlled laboratory conditions under different voltages; Researchers use this device to evaluate the partial discharges localization
on the voltage phase. The measurement can be performed using the same setup.
The oscilloscope is connected the phase of the voltage used for the experiment in
order to observe the distribution of impulses in the observation time (Fig. 3.3).

According to Figs. 3.2 and 3.3, it can be seen that the setup gives the expected
results since there are enough samples to observe the damped oscillating waveform
of impulses. The time-window is large enough to observe groups of impulses that

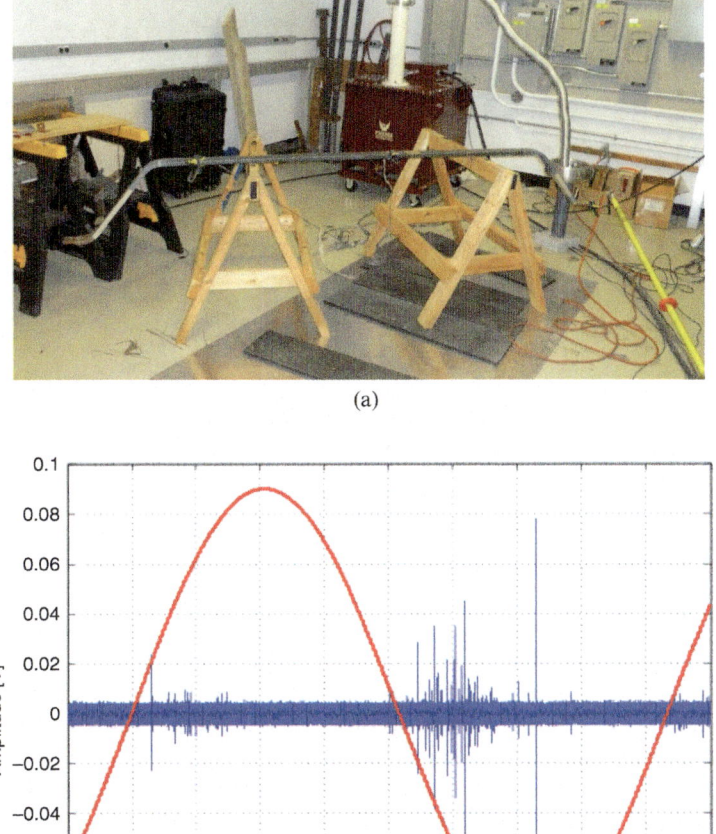

(a)

(b)

Fig. 3.3 Impulsive noise measured from a generator bar. (**a**) Impulsive noise source: the generator bar under 18 kV. (**b**) Impulsive noise (blue) coming from the generator bar and 60 Hz cycle (red) of the power supply

seem to be synchronized with the 60 Hz frequency of the voltage feeding the devices (Tesla-coil and generator bar). Since for each test in laboratory only one source of impulsive noise is used, it is interesting to verify that observation is also valid in several substations where many impulsive noise sources can take place [116].

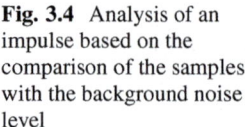

Fig. 3.4 Analysis of an impulse based on the comparison of the samples with the background noise level

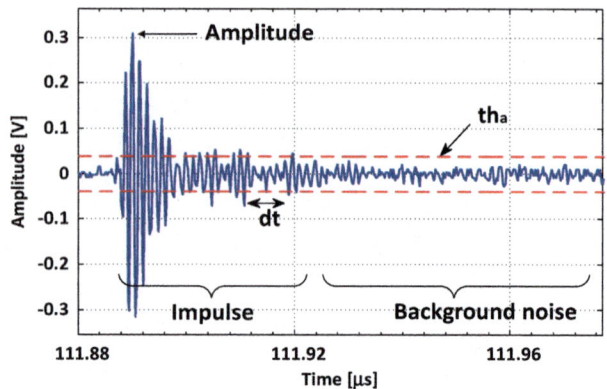

3.2.3   Impulse Detection Method

The measurement setup can provide recordings of impulsive noise over a large time window, but it is not possible to extract information concerning the impulses only. The extraction of these impulse parameters can be achieved through a method that allows to separate the impulses from the background noise in order to provide statistical impulse information. To do so, an impulse detection method can be designed in order to classify all measured samples as either background noise or an impulse event. The background noise is very limited in sample absolute value not exceeding a threshold denoted as th_a (Fig. 3.4).

A first step is to find the threshold th_a that separates the impulse samples from the background noise. All samples with an absolute value above this threshold are deemed to belong to an impulse; however some noise samples within an impulse might take values below th_a (Fig. 3.4); therefore the samples that belong to an impulse and that are also under the threshold th_a must be considered by our method of impulse detection. The part of an impulse that remains under the threshold th_a is composed of several consecutive samples, similar to the background noise between two impulses, and the second step of the impulse detection would determine whether those consecutive samples belong to an impulse or not; this decision will be based on the number of these consecutive samples compared to a duration threshold th_d to be defined.

The impulse detection is proceeded as follows:

- Determine a threshold th_a that separates the background noise from the impulses (Fig. 3.4).
- Define the variable dt as the duration between two consecutive samples above the threshold th_a.
- Collect the sequence $\{u_i\}$, which is the observation, in samples, of dt from the measurements.

- Determine the threshold th_d and for any element of $\{u_i\}$ above th_d, by considering that the time duration corresponding to that value u_i corresponds to a period between two different impulses.
- Once all the elements of $\{u_i\}$ are analyzed, separate the impulses from the background noise and extract the information about the amplitude, the duration and the IAT of the impulses.

The three different steps will be described in detail below. Regarding the calculation of the thresholds th_a and th_d, it is preferred to use prior information as little as possible because the system must be coping with the environment autonomously.

To find the threshold th_a that separates the impulse samples from the background noise, the noise recording is analyzed under an assumption of a simpler noise model. In order to simplify the calculation of th_a, we may consider that the samples of the measured noise are distributed according to a Bernoulli-Gaussian process; therefore, it can be assumed that the samples are distributed according to the pdf $f(z|\lambda, \sigma_0^2, \sigma_1^2)$:

$$f(z|\lambda, \sigma_0^2, \sigma_1^2) = \frac{\lambda}{\sqrt{2\pi\sigma_0^2}} \exp(-\frac{z^2}{2\sigma_0^2}) + \frac{1-\lambda}{\sqrt{2\pi\sigma_1^2}} \exp(-\frac{z^2}{2\sigma_1^2}), \qquad (3.1)$$

where σ_0^2 is the background noise variance, σ_1^2 is the impulses samples variance and λ is the probability to be in background noise. By applying the method of moments [74], we calculate the first three even sample moments m_2, m_4 and m_6 and solve the system of equations where the sample moments are equal to the statistical moments to find the three parameters σ_0^2, σ_1^2 and λ. $\langle . \rangle$ is defined as the mean function. In such a condition, the three first even moments is calculated as follows:

$$\begin{cases} m_2 = \langle z^2 \rangle = \lambda\sigma_0^2 + (1-\lambda)\sigma_1^2 \\ m_4 = \langle z^4 \rangle = 3[\lambda(\sigma_0^2)^2 + (1-\lambda)(\sigma_1^2)^2] \\ m_6 = \langle z^6 \rangle = 15[\lambda(\sigma_0^2)^3 + (1-\lambda)(\sigma_1^2)^3] \end{cases} \qquad (3.2)$$

The solutions of the system of equations $\langle z^i \rangle = m_i$, $i \in \{2, 4, 6\}$, give the following background noise variance σ_0^2 :

$$\sigma_0^2 = \frac{\alpha + \beta}{\gamma} \qquad (3.3)$$

where

$$\begin{cases} \alpha = 15m_2m_4 - 3m_6 \\ \beta = \sqrt{75m_4^2(4m_4 - 9m_2^2) + 270m_2(2m_2^2 - m_4)m_6 + 9m_6^2} \\ \gamma = 90m_2^2 - 30m_4 \end{cases}$$

With the variance estimated, the background noise level can be set using the universal threshold [117] $\sigma\sqrt{2\log n}$, with σ being the standard deviation of a zero-mean Gaussian sequence of length n. When generating samples from a Gaussian distribution, we can expect the samples not exceeding a specific value for a given number of samples and for a given variance of the Gaussian distribution; this specific value is called the universal threshold. A sequence of background noise samples of length n would correspond to an average inter-impulse time (IIT) and according to several measurements, an average IIT has a value between 10^5 and 10^6 samples, with a sampling frequency of 5 Giga-samples per second. Hence a background noise threshold defined as $th_a = 5\,\sigma_0$ and all samples above this threshold are considered to be part of an impulse.

Since impulsive noise is generated by partial discharges and Corona effects, an impulsive waveform can start and end with a damped oscillating waveform for a short duration (in order of μs). The impulse duration depends on several factors, which are the insulator nature where discharges occur, the voltage applied to conductors, the ambient humidity and the gap between the conductors. We consider that an impulse lasts as long as we can observe its amplitude above th_a for a duration remaining of the order of the partial discharge and Corona effect events. All samples above th_a in amplitude are considered to be part of impulsive events. The time lags between consecutive samples with amplitudes above th_a are compiled into a sequence of integers $\{u_i\}$ (Fig. 3.5). We use the ordered statistics of this sequence to determine the threshold th_d, which is the value of $\{u_i\}$ above which two large samples (i.e. with amplitude larger than th_a) are considered to be part of two distinct impulsive events. While observing the distribution of the $\{u_i\}$ values (Fig. 3.6), it is visually observed that the small values (usually inferior to 100) correspond to time lags between samples within the same impulse, while the values of $\{u_i\}$ corresponding to IITs are larger (between 100 and 1 million samples) and correspond to smaller values of the PDF. From this observation, we assume that the threshold th_d is the smallest value of $\{u_i\}$ for which the pdf (or more accurately the normalized histogram, as in Fig. 3.6) is negligible. Mathematically, this procedure can be described as follows: we denote by $\{v_j\}$, $1 \leq j \leq N$, the N unique values of the sequence $\{u_i\}$, sorted in increasing values. The equation below shows how the threshold th_d can be selected from the sequence $\{v_j\}$:

$$\underbrace{v_1 \ldots v_k}_{\text{within an impulse}} \quad \overbrace{\underbrace{v_{k+1} \ldots v_N}_{\text{IIT}}}^{th_d} ,$$

with $th_d = v_{k+1} - 1$, and $v_{k+1} = min_{1 \leq l \leq N-1}\{v_{l+1} : v_{l+1} > v_l + 1\}$.

After extracting the impulses, one can find their amplitudes and their duration; moreover, a signal composed of a succession of impulses can be created, and one can calculate its power spectrum.

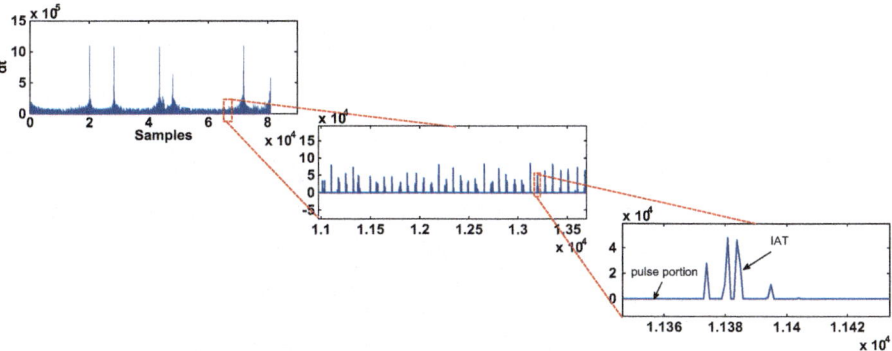

Fig. 3.5 $\{u_i\}$ sequence calculated from measurements

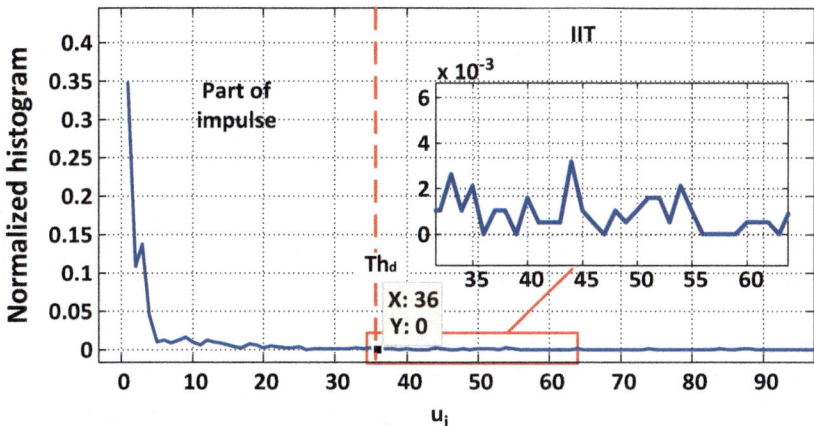

Fig. 3.6 Normalized histogram of the u_i

3.3   Measurements in Substation 1

3.3.1   Substation Presentation

Substation 1 is an air-insulated substation located in Quebec, where different types of equipment can be found (transformers, circuit-breakers, disconnecters, busbars, etc.). It also offers high voltage diversity by gathering pieces of equipment working under different voltage; areas by equipment voltage can be classified among possible values 230, 315 and 735 kV.

During the campaign, an antenna is placed in various locations, and it is pointed in the direction of HV equipments such as power lines, transformers and circuit-breakers located in the different voltage areas. For different antenna orientations, the noise is recorded during a 51.2 ms time window. This procedure is repeated on different days, in order to ensure diversity in the collected samples.

Fig. 3.7 735 kV equipment in Substation 1: switch gear (*Courtesy of Hydro-Québec*, 2014)

Sample diversity is the major preoccupation during impulsive noise measurement because the noise source locations are unknown and evolve with time. The impulses' observation is mostly due to partial discharge sources emitting directly to the antenna, but other impulses emanate from distant sources and are reflected by metallic structures. This environment is the best opportunity to capture how the real communication channel looks like.

The 735 kV area receives the voltage coming through power lines and redirects the electricity to the rest of the substation (230 and 315 kV). This zone mainly contains power lines, buses, circuit breakers and switch gears (Figs. 3.7 and 3.8). The 735 kV area certainly creates the most powerful impulsive noise recorded during our measurement campaign. "Powerful" impulsive noise refers to samples which are significant compared to the background noise samples. Significant impulsive noise consists of several impulses with a large amplitude. The vehicle was located 20–30 m from the closest 735 kV equipment (bars in Fig. 3.8). For any weather condition acoustic waves can be heard. This is generated by the phenomenon being the source of the corona noise coming from the power lines. The dry condition has been observed that leads to high impulsive noise. It can be conjectured are that humidity might increase the generation rate of the impulses but significantly attenuate their power.

The 315 kV area in Substation 1 is mostly composed of transformers. The measurement setup and the antenna are located 20 m from the equipment and the antenna is rotated in the direction of the transformers and the bushings (Fig. 3.9).

Due to the disposition of the pieces of equipment within Substation 1 (Fig. 3.10), some area access are limited. The equipment is located at the border of the substation, close to the fence and the location where the measurements are performed is not always connected to the rest of the substation (opened circuit-breaker).

Fig. 3.8 Overview of 735 kV area in Substation 1 (*Courtesy of Hydro-Québec*, 2014)

Fig. 3.9 Transformers: 315 kV area in Substation 1 (*Courtesy of Hydro-Québec*, 2014)

3.3.2   Locations of the Antenna

The oscilloscope being plugged to a generator inside the vehicle, in order to get closer to a specific voltage area, the antenna is surrounded by specific pieces of equipment. The antenna is oriented with three different positions that depend on the equipment. The bushing, circuit breakers and specific parts of the transformer such as termination between the transformer and the power lines (Fig. 3.11) are focused. The antenna can rotate 360° horizontally and 180° vertically. For substation automation, IEDs would be placed very close to power equipment, even in contact. The location of the antenna does not provide an exact observation of impulsive

Fig. 3.10 Overview of 230 kV area in Substation 1 (*Courtesy of Hydro-Québec*, 2014)

Fig. 3.11 Antenna orientation in Substation 1 (*Courtesy of Hydro-Québec*, 2014)

noise where the receivers would be placed. However, with the use of an LNA, the measured signal is amplified in order to compensate the attenuation caused by the distance between the antenna and the power equipment.

3.3.3   Results

In Fig. 3.12, it can be seen that the waveform of measured impulsive noise for the 230, 315 and 735 kV with the power spectrum of the impulses. We note that the measured impulsive noise is similar in the three areas of voltage in terms of impulses generation. A possible correlation between the frequency of the voltage (60 Hz) and the impulse generation is observable. In the next chapter, it can be seen that this possible correlation is not always observable. The impulses in the 735 kV area seem

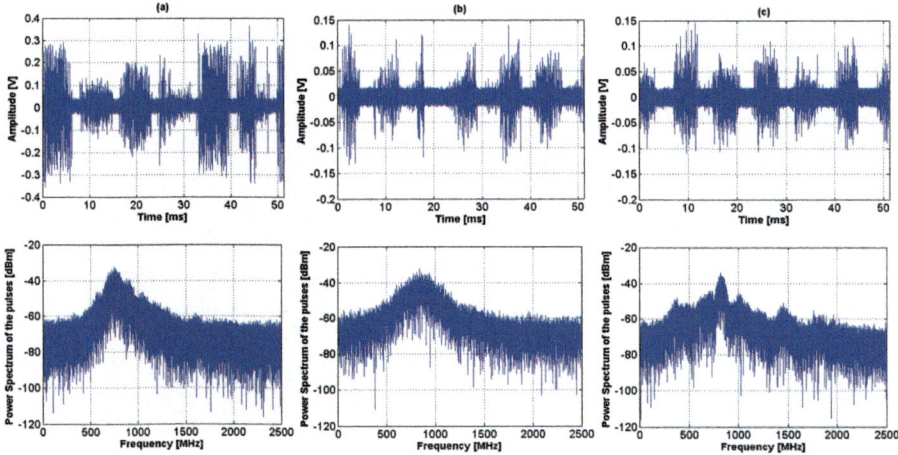

Fig. 3.12 Impulsive noise measurements in substation 1 for 735 kV (**a**), 315 kV (**b**) and 230 kV (**c**) areas; time domain (top) and power spectrum of all the impulses (bottom)

to be more powerful in terms of the number and the amplitude than the impulses measured in the other voltage areas. Section 3.5 studies the statistical characteristics of impulsive noise for the different range of equipment voltage.

3.4   Measurements in Substation 2

3.4.1   Substation Presentation

Substation 2 has a part located indoor for the 25 kV equipment. The 315 kV yard and the 25 kV area are connected through power transformers. The secondary part of the transformer (Fig. 3.13) is located indoor and allocates the electricity through the other pieces of equipment.

The 25 kV area is mainly composed of pieces of equipment listed below:

- Transformer (Fig. 3.13a)
- Breaker (Fig. 3.13b)
- Circuit-breaker
- Cable termination

The measurement setup was moved to different locations using a trolley (Fig. 3.13b) while the oscilloscope is plugged to the grid inside the substation. The 25 kV substation has the weakest impulsive noise recorded among our measurements in terms of amplitude and repetition rate.

Fig. 3.13 Substation 2, indoor area (*Courtesy of Hydro-Québec*, 2014). (**a**) Secondary transformer (25 kV) in Substation 2. (**b**) Measurement of impulsive noise coming from breakers in Substation 2

(a)

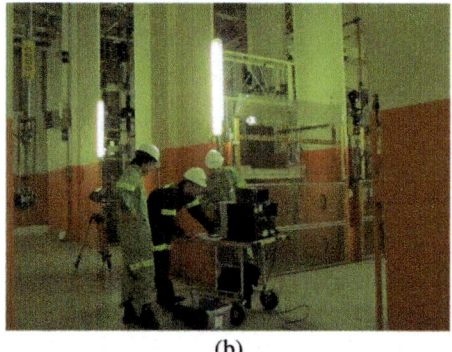

(b)

3.4.2   Locations of the Antenna

The antenna is oriented toward the secondary part of a 315–25 kV transformer, the secondary part being inside the building (Fig. 3.13a). We have also measured impulsive noise at other locations, such as breakers and buses. The measurements in the yard have been performed as in substation 1, which allows us to combine measurements from both substations in order to evaluate the noise characteristics for 315 kV equipment.

3.4.3   Results

The measurements in the 315 kV backyard are very similar to the results observed from substation 1 in terms of waveform of the noise and amplitude of the impulses. According to measurements in the 25 kV area (Fig. 3.14), impulsive noise in

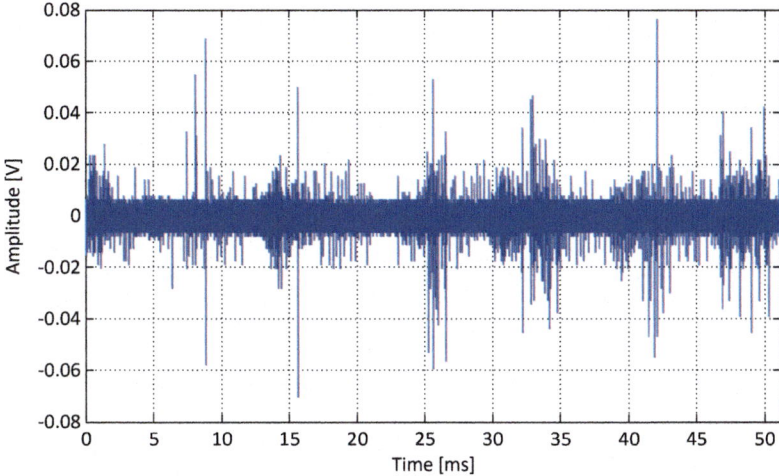

Fig. 3.14 Impulsive noise in the 25 kV indoor area of substation 2

Substation 2 seems weaker than the noise recorded outdoors; this is due to the fact that this part of the substation has the lowest voltage of equipment. The amplitude of the impulses and the repetition rate are lower in 25 kV area than in the vicinity of other equipment voltages. Although the presence of metallic structures in enclosed space, the 25 kV area seems to be a less harsh environment for wireless communications.

3.5   Classification of Impulsive Noise Characteristics

Substations 1 and 2 measurement campaigns have provided enough samples of impulsive noise to calculate representative characteristics of impulsive noise based on the equipment voltage. For the 315 kV area, we have used the impulses measured in substations 1 and 2 without any distinction. All the values presented are estimated using sampled mean and sampled variance derived from the characteristics observation.

3.5.1   Amplitude

The amplitude of the samples is a particular characteristic that is introduced as a criterion to evaluate the level that the samples can reach. We call amplitude of an impulse the largest absolute sample value within an impulse period. It is noted that increasing equipment voltage leads to an increase of the average

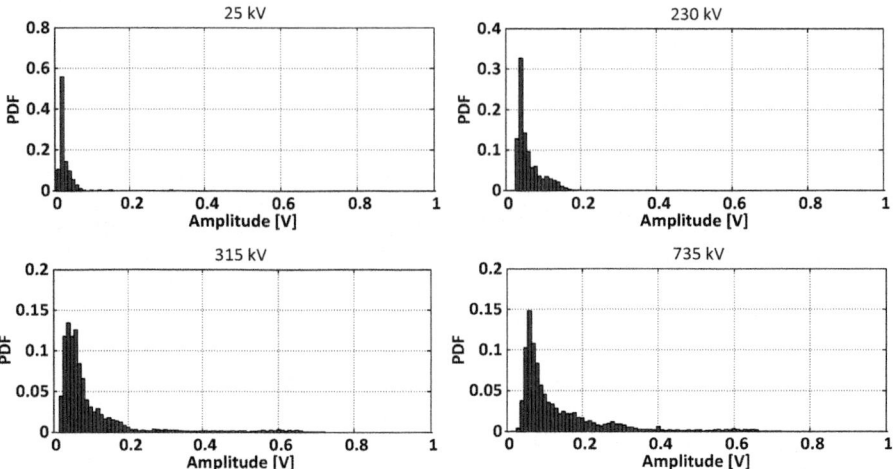

Fig. 3.15 Distribution of the impulse amplitude for 25, 230, 315 and 735 kV areas

Table 3.1 Impulsive noise characteristics

Impulsive noise characteristics	Voltage area			
	25 kV	230 kV	315 kV	735 kV
Amplitude (V)	0.026	0.061	0.098	0.136
Impulse duration (samples)	132	72	70.39	61.54
Repetition rate (impulses/second)	6.372×10^3	1.278×10^4	1.21×10^4	1.635×10^4
Impulse sample variance	1.456×10^{-4}	8.115×10^{-4}	4.13×10^{-3}	5.702×10^{-3}
Background noise variance	4.207×10^{-6}	2.041×10^{-5}	2.981×10^{-5}	3.46×10^{-5}
Total noise variance	4.407×10^{-6}	2.1×10^{-5}	3.084×10^{-5}	3.81×10^{-5}

impulsive noise amplitude (Fig. 3.15). The impulses have a larger amplitude when the voltage increases. Average amplitude results from the measurement campaigns are displayed in Table 3.1; It is confirmed that higher voltage substation generate a more powerful impulsive noise in terms of amplitude of the impulses.

3.5.2   Impulse Duration

The impulse duration is defined as the number of samples between the first and the last sample of an impulse. In terms of time, the impulse duration can be defined by the following equation:

$$\text{Impulse duration} = \frac{\text{Number of samples in an impulse}}{\text{Sampling frequency}} \qquad (3.4)$$

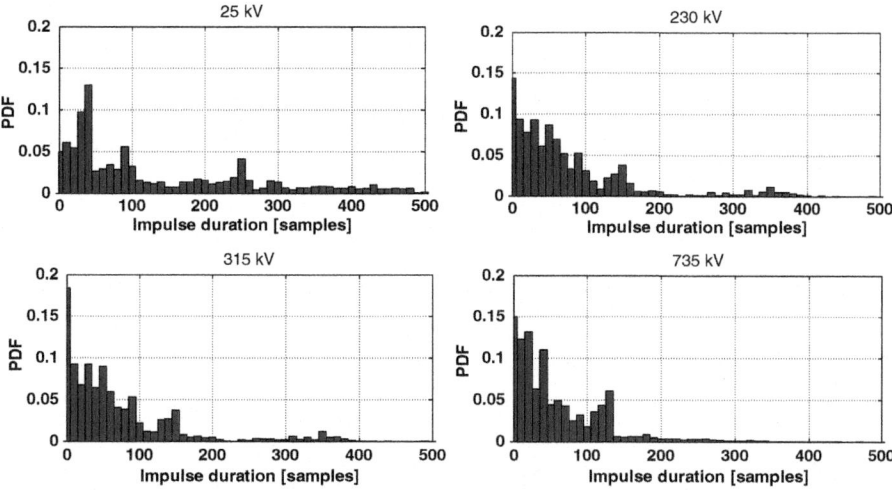

Fig. 3.16 Distribution of the impulse duration for 25, 230, 315 and 735 kV areas

The impulse detection method provides the impulse duration for the four ranges of substation voltage (Fig. 3.16). To see if the amplitudes and the impulse durations are correlated, we use the data extracted from the observation to calculate the Pearson's correlation coefficient ρ:

$$\rho = \frac{cov(\text{amplitude, impulse duration})}{\sigma_{\text{amplitude}} \times \sigma_{\text{impulse duration}}}, \tag{3.5}$$

where, cov denotes the covariance, $\sigma_{\text{impulse duration}}$ and $\sigma_{\text{amplitude}}$ the standard deviations of the impulse durations and amplitudes, respectively. For a correlation coefficient equal to 1, the data would be perfectly correlated. Using the observation data and (3.5), the correlation coefficient is 0.825, so there is a high correlation between the amplitudes and the impulse durations. We find similar results on different sequences of noise; it can be concluded that, for a given noise sequence, the larger the amplitude of impulses, the longer they last.

It is quite apparent that the impulses approximatively last the same duration, on average, no matter what voltage is applied to the equipment (Table 3.1). While the amplitude increases with the equipment voltage, the impulse duration remains stable; this observation can be easily explained by the impulse detection method. Impulses are detected based on the evaluation of the sample value compared to the background noise level. However both the amplitude of the impulses and the background noise level increase with the voltage of the equipment. Although in theory a large impulse lasts longer than a small impulse, a higher background noise level would hide the last samples of the impulse and one would observe a shorter duration.

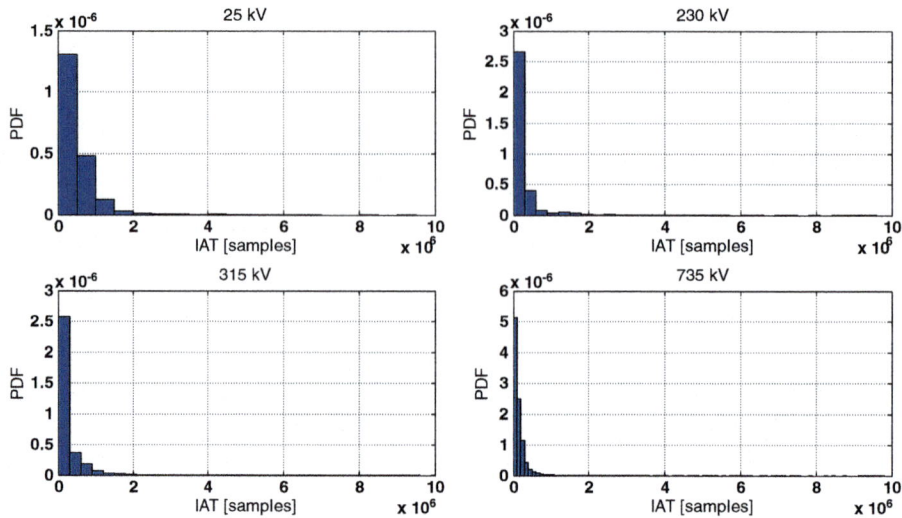

Fig. 3.17 Distribution of the IAT for 25, 230, 315 and 735 kV areas

3.5.3   Repetition Rate

The repetition rate can be estimated by the number of impulses by second or by the inter-arrival time, which is the duration between two consecutive impulse generations. The IAT is calculated from the impulse detection method as the duration between the first sample of two consecutive impulses. We observe from Fig. 3.17 that the IAT has a distribution.

Regarding the number of impulses, we observe from the distribution but also from the average repetition rate in Table 3.1 that when the substation voltage increases, the number of impulses per second increases too. More impulses are present in substations with a higher equipment voltage.

3.5.4   Sample Value

The time domain waveform and sample statistics can be used to evaluate the accuracy of a model. Although most of comparisons between models use histogram of impulsive noise samples without any distinction of whether the samples belong to the background noise or the impulses, the impulse samples will be separated from the background noise. Thus, their distribution will be studied. This decision is justified by pointing out that measurements provide impulsive noise where the background noise samples are predominant (in terms of number of samples) compared to the impulse samples. In Fig. 3.18, we observe that the impulses do not follow a single Gaussian distribution, but they seem to be distributed according to

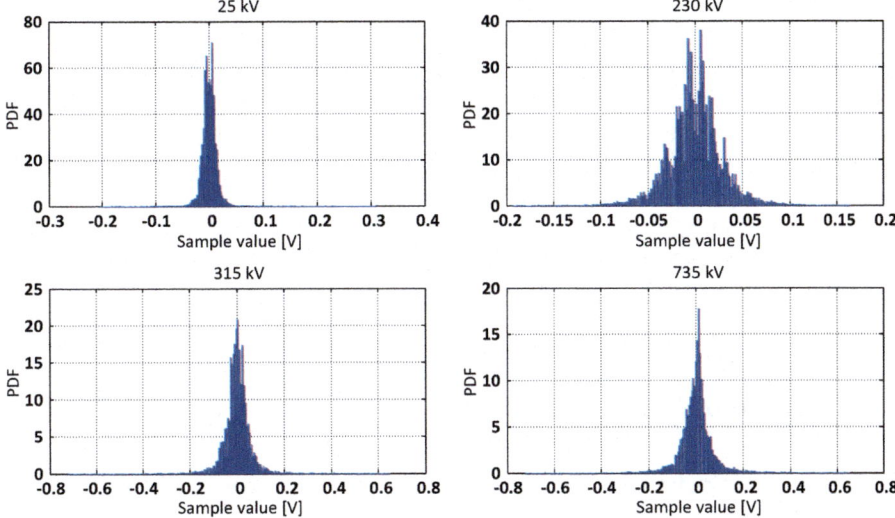

Fig. 3.18 Distribution of the sample value for 25, 230, 315 and 735 kV areas

Gaussian mixture. Table 3.1 shows that when the equipment voltage increases, the variance of the impulse samples increases too. This observation totally makes sense, since the variance of the impulse samples and the impulse amplitude are correlated; as a reminder, the noted amplitude is the maximum of absolute value generated by the distribution of impulse sample.

3.5.5 Representative Characteristics

Representative characteristics of impulsive noise are presented in Table 3.1 to provide an overview of the RF impulsive noise in all high voltage areas. For the impulses, the amplitude, the repetition rate and the impulse duration are average values. Some information about the impulses (Table 3.1) can be used directly to estimate the parameters of the Class-A and 2-state Markov chain models.

3.6   An Experimental Characterization of the Discharge Sources

A non-invasive PD measurement using wideband antenna is proposed for EMIs induced by PDs in substations. Signal processing tools using short-time analysis are presented for a full PD characterization by their electromagnetic radiations from which first-order and second-order statistics can be derived. These can be implemented into any wireless electronic devices using antennas for a rapid and on-line PD diagnostic in HV equipment.

These signal processing tools are also valuable for the characterization of wireless communication channels in the presence of impulsive noise in substations. One remarkable characteristic of PD is that the rise-time is very rapid so that the spectrum can span to 3 GHz, which can overlap communication systems operating at 2.4 GHz. Indeed, researchers have shown experimentally that performances of conventional wireless communication systems using 2.4 GHz ISM bands are degraded by PDs [71, 72, 114, 118, 119].

3.6.1 Amplitude of Measured Signals

Electromagnetic radiations generated by PD sources are characterized by any RF gain in the measurement setup by removing the antenna factor. As a result, the amplitude of the measured signal in voltage (V) is converted into electric field strength in (V/m).

We denote $u(\theta, t)$ as the impulsive noise waveform measured in electric field strength (V/m), where θ is a set of random variables characterizing its duration, occurrence and other physical parameters. By denoting $u_m(\theta, t)$ as the impulsive waveform measured in voltage (V), we have:

$$u(\theta, t) = u_m(\theta, t) \sqrt{\frac{Z_0 4\pi}{R G_r \lambda^2}} \tag{3.6}$$

where R is the load resistance and G_r is the gain of the RF system including the filter, the RF amplifier and the antenna. $Z_0 = 120\pi$ Ω is the freespace impedance and λ is the wavelength of wideband antenna. Note that this relationship is valid in far-field conditions. In practice, PD sources and the antenna are separated by several meters, thereby establishing far-field conditions.

3.6.2 Signal Processing Tools for Impulsive Noise Measurement

3.6.2.1 The Denoising Process

In practice, the measured signals consist of overall background noise produced by thermal noise in RF components, interleaving artefacts, clock feedthrough noise in the scope and RF signals which come from cellular phones at 1.7 and 1.9 GHz. The resulting noise received by the antenna is written as:

$$x(\theta, t) = \sum_k u_k(\theta_k, t) + n(t) \tag{3.7}$$

where $n(t)$ is the signal of the overall background noise. Over a long observation time, the presence of many impulses is a superposition of impulsive transient waveforms, where the occurrences are randomly distributed over this interval of time.

By using a denoising process, the overall background noise can be removed to ensure a better temporal and frequency location of each impulse. The wavelet transform is used to decompose the signal into components over which a threshold has been applied to remove low wavelet coefficients. The wavelet transform is used to decompose a signal on a wavelet orthonormal basis. It defines a multi-resolution representation of the signal [80].

The discrete wavelet transform is based on the convolution of the signal with a pair of quadrature mirror filters. The signal is decomposed by these filters successively. Since impulsive components are above the background noise component, their wavelet coefficients are higher than the background noise [80]. Low coefficient values can be set to zero by a hard threshold defined by Donoho [120] such that the threshold T_h is given by:

$$T_h = \sigma_i \sqrt{2\log(M_i)} \tag{3.8}$$

where σ_i^2 is the variance of the background noise and M_i is the number of samples for a given level of decomposition i. Studies reported in [120, 121] have shown that the background noise standard deviation is estimated by:

$$\sigma_i = \frac{\text{Med}_i}{0.6745} \tag{3.9}$$

where Med_i is the estimated median value of the signal. Next, impulses are extracted by rescaling the threshold value at each decomposition level, and the reconstructed signal is obtained by the inverse of the wavelet transform. The denoising process can be summarized by Fig. 3.19. The wavelet decomposition is applied to measurements $x(\theta, t)$, where low wavelet coefficient values are set to zero by hard threshold T_h at each level of decomposition. Impulsive components can be estimated by using the inverse of wavelet transform. An extensive analysis on denoising PD signals shows that a suitable estimation of impulsive transient-noise waveforms can be performed using Daubechies wavelets with 8 vanishing moments for 30 levels of decomposition [122–124]. These specific parameters are valid for the specific sampling frequency.

3.6.2.2 Short-Time Analysis for Impulsive Signals

Over a long observation time, many impulses can be observed in measurements with different time occurrences. When their amplitude is significant, it is reasonable to say that the process is non-stationary. In such instances, it may be useful to conduct a short-time analysis to preserve information regarding the time-frequency location of impulse events.

Fig. 3.19 Denoising process
using wavelets

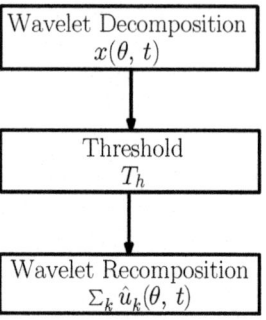

Spectrograms can be used to yield a time-frequency representation of measured signals. Based on the short-time Fourier transform, impulses can be written as follows:

$$U(\theta, t_w, f) = \int_{-\infty}^{+\infty} u(\theta, t) w(t - t_w) e^{-j2\pi ft} dt \qquad (3.10)$$

where $w(t)$ is a square-integrable temporal window function with a length of t_w. The spectrogram can be computed by the squared magnitude of the STFT as:

$$S_{uu}(\theta, t_w, f) = \frac{1}{Z_0} |U(\theta, t_w, f)|^2 \qquad (3.11)$$

where each segment $S_{uu}(t_w, f)$ is the power density of the process located in (t_w, f), and $Z_0 = 120\pi$ is the impedance of free space. To ensure a better time-frequency localisation, the window function must be defined properly, as well as, the number of samples that overlap between adjoining sections and the number of FFT. These parameters will be defined in the next section.

3.6.2.3 Temporal Location of an Impulse

Short-time analysis allows for the study of first-order statistics in terms of power density, inter-arrival time and time occurrence.

The power density of PD sources can be estimated in a selected bandwidth Δf near a given resonant frequency f_0, as follows:

$$S_{uu}(\theta, t_w) = \frac{1}{Z_0} \int_{f_0 - \Delta f/2}^{f_0 + \Delta f/2} \frac{|U(\theta, t_w, f)|^2}{\langle w(t), w(t) \rangle} df \qquad (3.12)$$

where $\langle w(t), w(t) \rangle$ is defined by:

$$\langle w(t), w(t) \rangle = \int_{-\infty}^{+\infty} w(t) w^*(t) dt \qquad (3.13)$$

and $w^*(t)$ is the complex conjugate of $w(t)$.

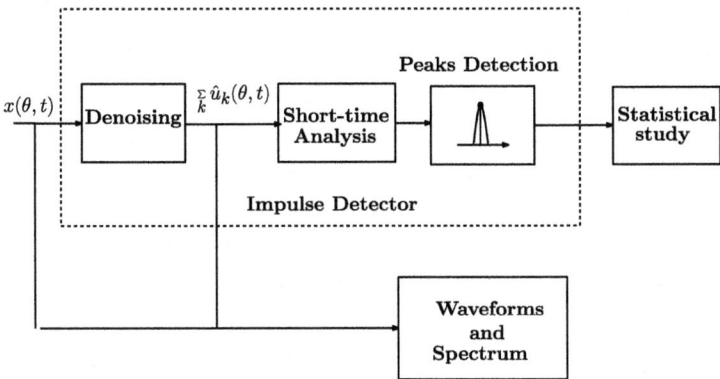

Fig. 3.20 Characterization process

An impulse detector is used for an experimental characterization of PD sources as illustrated in Fig. 3.20. First, impulsive noise emitted by PD sources is extracted from overall background noise via a denoising process. Next, the power density of each impulse is calculated using the short-time Fourier analysis. A peak detection is employed to measure the power density value at t_w, denoted as P_d. PD sources can be characterized in terms of their power spectral density.

The characterization metrics are defined based on the probability distributions of some physical parameters such as amplitude, inter-arrival time and occurrence (first-order statistics), as well as power spectral density and autocorrelation function (second-order statistics).

3.6.3   Characterization Based on First-Order Statistics

PD processes can be characterized by first-order statistics such as amplitude, inter-arrival time and occurrence. These parameters provide relevant information regarding the PD mechanism, its physical process and electrical aging of HV equipment. These statistical quantities are useful for the characterization of wireless channels in substations. Inspired by related research projects [21, 45, 124], first-order characterization metrics for the PD sources under AC voltages are defined in terms of:

- the power density of impulsive noise radiations P_d distribution (W/m^2);
- the discharge occurrence t_n distribution (s);
- the inter-arrival time Δt distribution (s).

The phase-resolved partial discharge (PRPD) representation is useful to show the power density distribution in a restricted interval time between t_{i-1} and t_i where $\Delta t_i = t_i - t_{i-1}$ [55]. PRPD is expressed as:

$$p(P_d|\Delta t_i) = \int_{t_{i-1}}^{t_i} p(t_n)p(P_d \mid vert t_n)dt_n \qquad (3.14)$$

where $p(t_n)$ is the probability per time unit of an impulse occurrence at $t_n \in$ $[t_{i-1}, t_i]$. $p(P_d|t_n)$ determines the probability of a discharge event having a power density P_d if its occurrence is t_n. The PRPD representation is widely used in the study and analysis of these discharge sources on insulation systems [56, 70, 125, 126]. Because these sources are typically generated by AC voltages, the processes are cyclostationary. Therefore, the PRPD representation can be used to characterize the occurrences and power densities of these EMI sources per unit time or phase.

The measurement data consist of a train of impulses randomly located in time. The spectrogram is used to detect each impulse for which power density, occurrence, and inter-arrival time can be calculated. For a suitable time-frequency localisation, we choose a Hamming window is used whose length is set according to the average duration of impulses.

3.6.3.1 PRPD Representation

EMI from PD can be characterized using PRPD representation, as depicted in Fig. 3.21. The color map indicates the number of discharges that had occurred. It is clear that PD characteristics can be described in terms of time-dependent random variables when PD activity can follow the cyclostationary process induced by AC voltages. The power density of these impulses are randomly distributed such that some PDs can reach 15 mW/m^2. Impulsive events occur randomly over time, even though they take place at every half-cycle of the AC voltage. Note that since the AC power-line of the scope has been used to capture impulses periodically, the measurement setup is not synchronized with AC voltages applied to PD sources. Thus, a phase shift is observed in the PRPD.

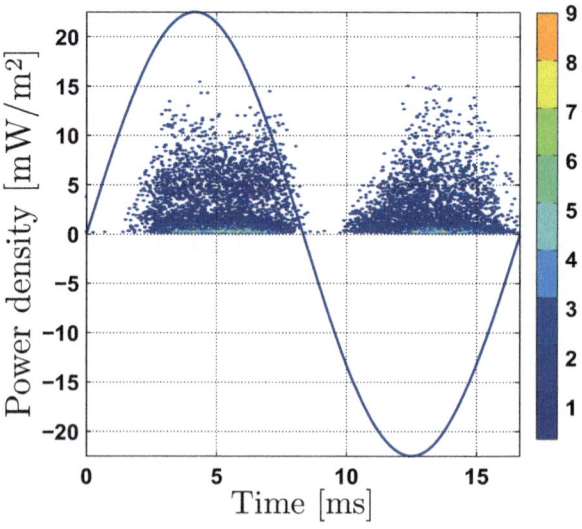

Fig. 3.21 PRPD of EMI from PD activity in a 735 kV substation

Table 3.2 Summary of statistical quantities

	Total PD	PD first half-cycle	PD second half-cycle
Rate (disch. per cycle)	408	235	129
Avg. power density (mW/m^2)	1.4655	0.7891	0.5637

The total PD rate is 408 discharges per cycle. The number of PD events is more predominant during the first half-cycle, i.e. within the time interval of 1–9.5 ms. Indeed, the PD rate is 235 discharges per cycle during the first half-cycle whereas the PD rate is 129 discharges per cycle during the second one. The average total power density is 1.4655 mW/m^2. The average power density during the first half-cycle is 0.7891 mW/m^2 and the second one is 0.5637 mW/m^2. These measured quantities are summarized in Table 3.2.

3.6.3.2 Statistical Distribution of PD Characteristics

PD characteristics can be presented in terms of statistical distributions, such as PDFs and complementary cumulative distribution functions (CCDFs) of power density, inter-arrival time and time occurrence. The CCDF is given by:

$$\bar{F}_X(x) = 1 - \int_{-\infty}^{x} f_X(\tau)d\tau \tag{3.15}$$

where f_X is the PDF of the studied random variable produced by PD activity. PDFs and CCDFs are depicted in Figs. 3.22, 3.23, and 3.24. Power density and IAT distributions follow power law distributions with a fast decay (see Figs. 3.22 and 3.23). The average power density is $\bar{P}_d = 1.4655$ mW/m^2 and the average IAT is $\bar{\Delta t} = 47\,\mu$s. PD occurrence distribution within a cycle has two separate Gaussian distributions. Under AC voltage, PD discharges occur at every half-cycle because the critical value for a discharge is reached when the local electric field in the PD site is intense. This critical value and the intensity of the electric field affect amplitude, IAT and occurrence of PD processes. On average, the majority of discharges take place at 6.2 ms during the first half-cycle of the AC voltage and at 12.56 ms during the second one. Such events follow a cyclostationary process due to AC voltages.

3.6.4 Characterization Based on Second-Order Statistics

Second-order statistics can be useful in characterizing RF signals from PD activity. These allow for the analysis of the statistical distribution of an EMI's power density over frequency, and also facilitates the identification of signal spectral characteristics induced by PD. These second-order statistics are defined by power spectral density (PSD) and autocorrelation function (ACF). It might be difficult

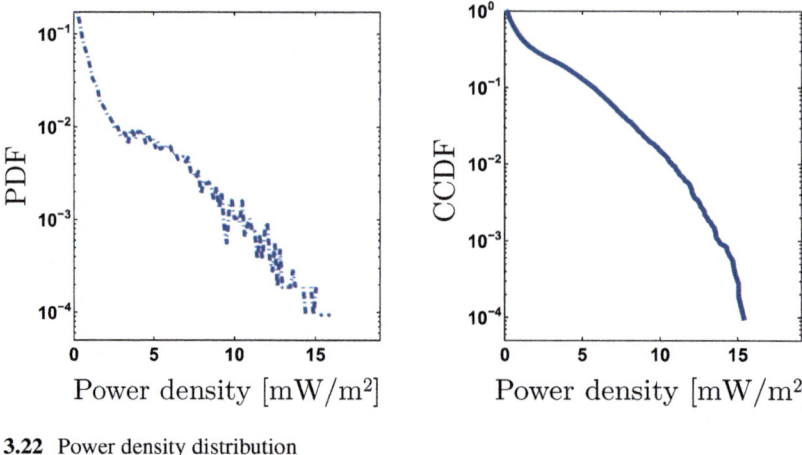

Fig. 3.22 Power density distribution

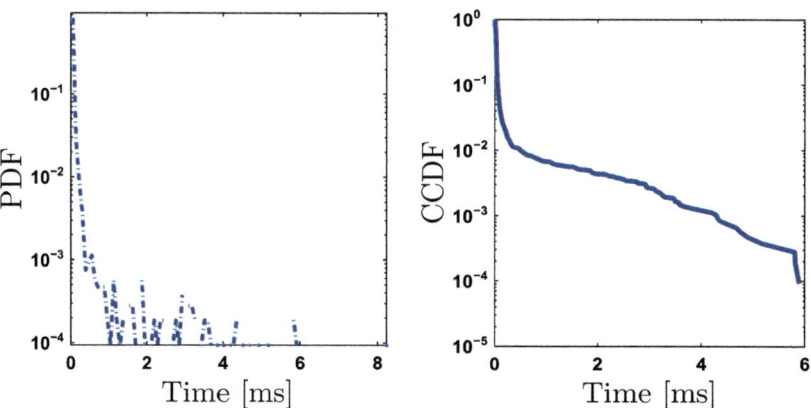

Fig. 3.23 Inter-arrival time distribution

to characterize non-stationary random processes of real-valued functions using second-order statistics. Nevertheless, PSD and ACF can be estimated using numerical methods. By taking measurements as a discrete-time signal $u(n)$, the relationship between the time variables t and n is given by:

$$t = nT_s \qquad (3.16)$$

where T_s is the sampling period. Moreover, we assume that a characterization using second-order statistics is restricted to a single impulse delimited by its duration. The length of the measured discrete-time signal is given by M. As a result, an estimation of PSD and sample ACF can be provided. The estimated PSD can be obtained by using the periodogram via discrete Fourier transform. For an impulse $u(n)$ sampled at F_s samples per unit time, the estimated PSD is given by:

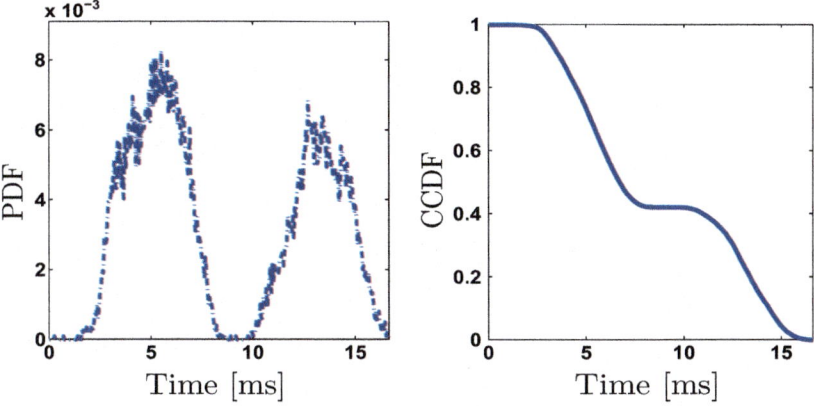

Fig. 3.24 Time occurrence distribution

$$P_{uu}(f_k) = \frac{1}{M} \left| \sum_{n=1}^{M} u(n) e^{-j2\pi n f_k} \right|^2 \tag{3.17}$$

where the frequency is given by $f_k = k/M$, where $k = \{0, 1, \cdots, M - 1\}$, and M is the number of observations.

The ACF can be found by taking the covariance of $u(n)$ and $u(n - i)$. The autocorrelation for a lag i as r_i is defined such that:

$$r_i = \frac{\sum\limits_{n=i+1}^{M} (u(n) - \bar{u})(u(n - i) - \bar{u})}{\sum\limits_{n=1}^{M} (u(n) - \bar{u})^2} \tag{3.18}$$

where $\bar{u}$ is the estimated mean value of the measured impulse. The ACF is used to check whether samples noises are correlated when an impulse is observed. Accordingly, it allows us to select an approach to modeling impulsive noise waveforms that fits measurements using a discrete-time filter.

3.6.4.1 Typical Waveform and Spectrogram

A typical waveform from PD activity measured in the substation is depicted in Fig. 3.25. In this case, the impulse has been measured at a sampling frequency of $F_s = 10$ G samples/s. The PSD, the ACF and the spectrogram are used to represent the RF signal. To have a suitable time-frequency resolution of the impulse, a Hamming window of $t_w = 3.2$ ns length is used. The FFT length is $N_{\text{fft}} = 512$ and 70% overlap is used.

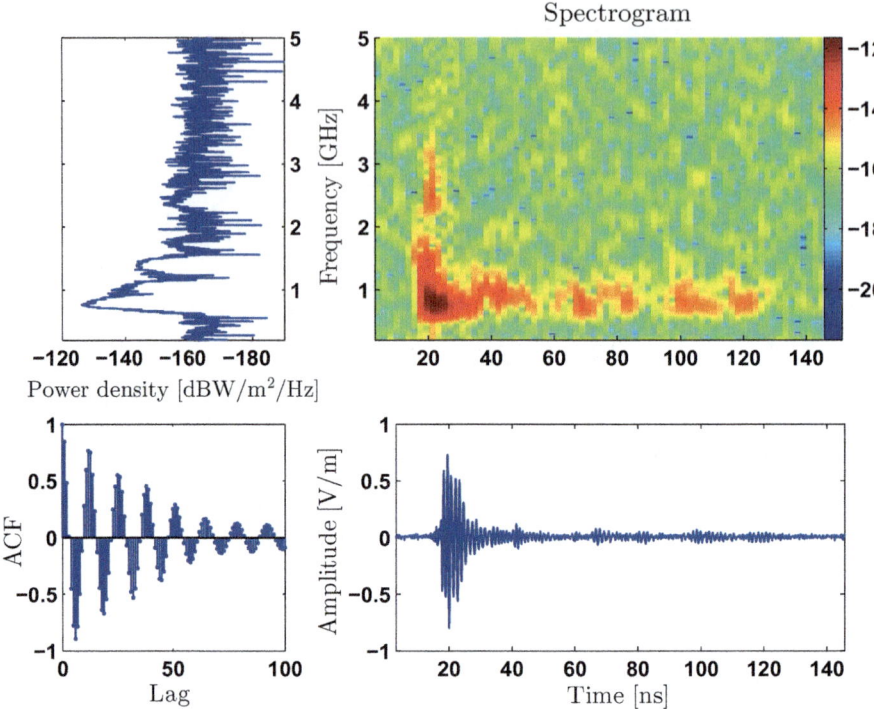

Fig. 3.25 Typical impulsive waveform

The waveform of a discharge is characterized by a short rise time (4 ns), a long fall time (50 ns) with damped oscillation around $f_0 = 800$ MHz, and high amplitude, maximal amplitude 0.75 V/m. During the rise time, the discharge can cover a large frequency range of 800 MHz to 3 GHz when the discharge amplitude is high. During the fall time, it covers a frequency range of 800 MHz to 1 GHz. The ACF exhibits a damped oscillation decay due to the damped oscillation and transient effect of the measured impulse.

An impulse generated by a PD can be approximated by a damped harmonic oscillator such that:

$$u(t) = u_0(e^{-at} - e^{-bt}) \sin(2\pi f_0 t + \varphi) \qquad (3.19)$$

where u_0 is the amplitude of the impulse, a and b are respectively rise time and fall time decay, f_0 is the resonant frequency and φ is the phase shift of the measured impulse.

The ACF of $u(t)$ in Eq. (3.19) can be approximated by:

$$R_{uu}(\tau) = \mathbb{E}\left[u(t)u(t+\tau)\right] \approx \frac{u_0^2}{2}(e^{-a\tau} + e^{-b\tau}) \cos(2\pi f_0 \tau) \qquad (3.20)$$

where τ is the time lag.

From these measurements, one can conclude that memoryless impulsive noise models can be limited. Indeed, these noise models generate impulses in one sample in which the resulting noise process is *iid*. Impulses from PD activity exhibit bursty behaviour and the ACF indicates that, in the presence of transient impulsive noise induced by PD, samples are highly correlated (see damped oscillated decay in the ACF).

3.6.4.2 Power Spectral Density

The spectral characteristics of PD impulses can be provided via second-order statistics. The periodogram is used to estimate the PSD of these impulses in this work.

3.6.4.3 Power Spectral Density of an Impulse

The PSD of measured PD impulse before and after the denoising process are depicted in Fig. 3.26. When the wavelet transform is applied to the measurements, low wavelet coefficients are considered to be ambient noise components. The hard threshold removes them and keeps high wavelet coefficients, i.e. impulsive components which yields a good estimation of the spectral characteristics of an PD impulse. Furthermore, deep fades can be observed at some frequencies. This might be due to the presence of multipath effects in which multiple reflections of EM waves are observed by the antenna.

3.6.4.4 Average Power Spectral Density

Based on the total number of $N_{\text{disch}} = 12,400$ measured discharges, the average PSD is estimated by using the periodogram in Eq. (3.17) as:

$$\bar{P}_{xx}(f) = \mathbb{E}\left[P_{xx}(\theta, f)\right]$$

$$= \frac{1}{N_{\text{disch}}} \sum_{l=1}^{N_{\text{disch}}} \frac{1}{M} \left| \sum_{n=1}^{M} x(\theta_l, n) e^{-j2\pi nf} \right|^2 \tag{3.21}$$

where $x(\theta_l, n)$ is raw or denoised data. M is the number of observations. This has to be larger than the duration of impulses. The average PSD is depicted in Fig. 3.27 where $M = 1024$ samples, i.e. 10.24 μs. In Fig. 3.27a, the average PSD is calculated from noisy data. The green curve is the measured ambient noise PSD. RF communications can be seen around 900 MHz and around 1.9 GHz. Harmonics at 1.25, 2.5 and 3.75 GHz are created by interleaving artefacts and clock feedthrough from the oscilloscope. On average, PD impulses can cover a frequency range of 800 MHz to 2 GHz.

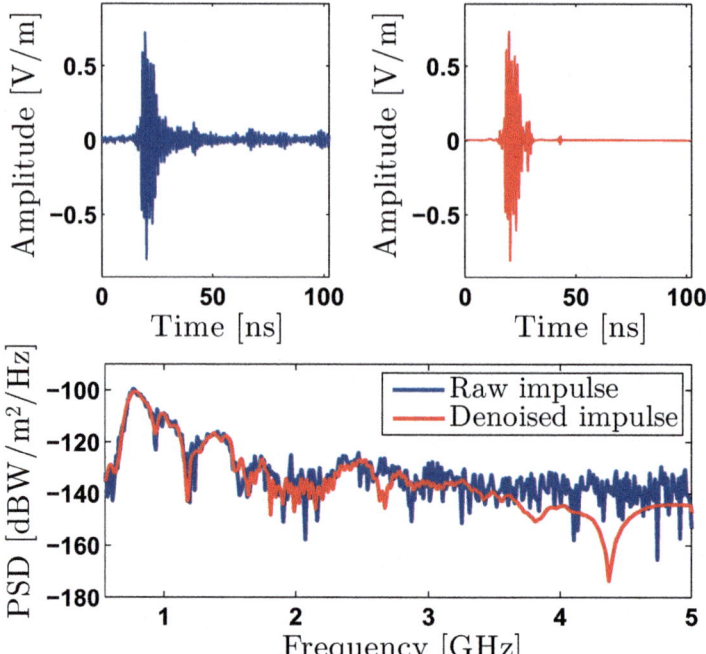

Fig. 3.26 Waveforms and PSD of an impulse

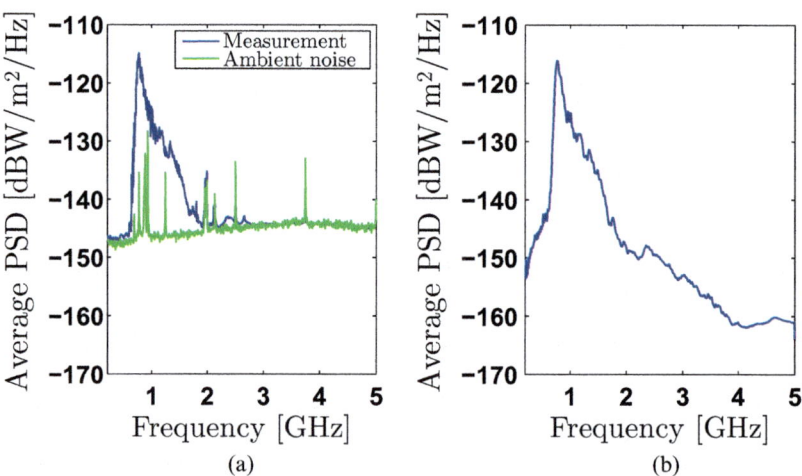

Fig. 3.27 Average power spectral density. (**a**) Raw PSD. (**b**) Denoised PSD

3.7 Representative Parameters for Classic Impulsive Noise Models

We have led several measurement campaigns over 2 years, which provides us with over 120 noise sequences of 256 millions of samples each. Applying the impulse detection method in Sect. 3.2.3 to the measurements allows us to estimate representative parameters for the existing impulsive noise models. In this section, the representative parameters are provided for the Middleton class-A model and the 2-state Markov chain.

3.7.1 Bernoulli-Gaussian Model

Two-state Markov chain model requires information about the impulse duration, the repetition rate, the variance of the background noise and the variance of the impulse samples. We use the same representation as in Sect. 2.4.3.2, state 0 represents the background noise and state 1 represents the impulse.

The impulse duration is represented by the probability to leave state 1 (p_{10} in Fig. 2.15) and the repetition rate is represented by the probability to transit from state 0 to state 1: p_{01}. The parameters are given in Table 3.3.

3.7.2 Middleton Class-A Model

We have mentioned in Chap. 2 that the method of moments does not guarantee satisfying results all the time. We propose to consider another method, based on the definition of the parameters, to calculate their value. By using the impulse detection method explained in Sect. 3.2.3, we can derive the following impulsive noise characteristics:

- The average impulse duration
- The repetition rate in impulses/second

Table 3.3 Representative parameters for substation impulsive noise modeled by two-state Markov chain (MC2)

MC2 parameters	Voltage area			
	25 kV	230 kV	315 kV	735 kV
p_{01}	1.274×10^{-6}	2.556×10^{-6}	2.316×10^{-6}	3.27×10^{-6}
p_{10}	7.55×10^{-3}	0.013	0.014	0.0162
σ_0^2	3.994×10^{-6}	2.041×10^{-5}	2.702×10^{-5}	3.46×10^{-5}
σ_1^2	1.456×10^{-4}	8.115×10^{-4}	3.246×10^{-3}	5.702×10^{-3}

Fig. 3.28 QQ-plot of the impulsive noise measured in substation 1 (X Quantiles) vs impulsive noise generated by the Class-A model (Y Quantiles) for two parameter estimation methods: method of moments and our impulse detection method in Sect. 3.2.3. (**a**) Method of moments; $A = 2.51 \times 10^{-4}$ and $\Gamma = 0.047$. (**b**) Method of impulse detection; $A = 0.001$ and $\Gamma = 0.0067$

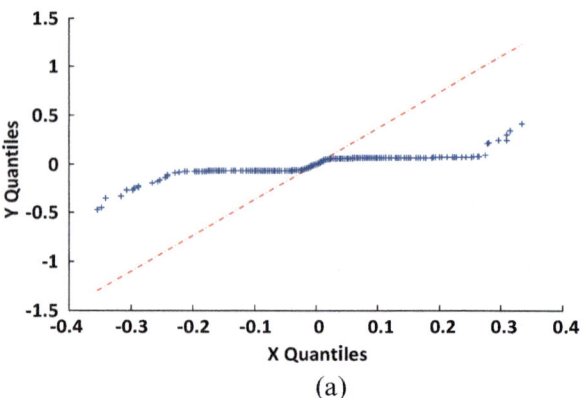

(a)

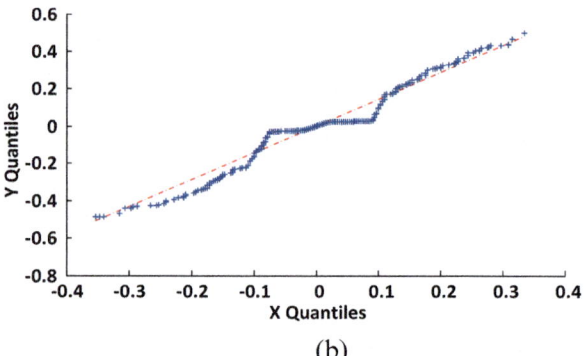

(b)

- The variance of the background noise
- The variance of the impulses
- The variance of the total noise (impulses + background noise)

After simulations, we compare the waveform of a sequence of measured impulsive noise with class-A noise parameterized using the method of moments and using the impulse detection method. The results are presented in the QQ (quantile-quantile)-plot graphs. These plot the quantiles of two probability distributions to be compared against each other [127]. In Fig. 3.28a, b, the two QQ-plot graphs confirm that the method of impulse detection provides better results for the samples distributions than using the method of moments.

In conclusion, the class-A model provides a canonical description of non-Gaussian noises for telecommunications and is applicable for general communication problems. The parameters estimation of the model is simple using the method of moments but hardly efficient for a direct application [74]; we prefer then to use the impulse detection method and calculate the parameters using their definition. This model gives good results to generate the amplitudes of impulsive noise, but without any time correlation.

Table 3.4 Representative parameters for substation impulsive noise modeled by Middleton class-A

Class-A parameters	Voltage area			
	25 kV	230 kV	315 kV	735 kV
A	1.668×10^{-4}	1.843×10^{-3}	1.705×10^{-4}	2.013×10^{-4}
Γ	2.89×10^{-2}	2.51×10^{-2}	7.21×10^{-3}	6.07×10^{-3}
σ	4.407×10^{-6}	2.1×10^{-5}	3.084×10^{-5}	3.81×10^{-5}

The class-A model uses three parameters that we have defined in Chap. 2 and we prefer to calculate the parameters using the average values of the impulse noise characteristics (Table 3.1). Given the parameter definition, we can calculate the following parameters:

$$A = \text{repetition rate} \times \text{average impulse duration} \qquad (3.22)$$

$$\Gamma = \frac{\text{background noise variance}}{\text{impulse sample variance}} \qquad (3.23)$$

$$\sigma^2 = \text{sampled variance(impulsive noise samples)} \qquad (3.24)$$

The parameter values are shown in Table 3.4

3.8 Conclusion

In this chapter, we have used a measurement setup that captures samples on a time window that is long enough to allow a statistical study of the noise. Signal processing tools using short-time analysis are presented for a full PD characterization by their electromagnetic radiations from which first-order and second-order statistics can be derived. This information can then be applied towards the development of rapid on-line remote monitoring and diagnostic tools in HV equipment, and/or for characterizing and modeling wireless channels in substations.

The appropriate sampling frequency highlights the correlation between the samples. The correlation is observable in terms of impulse duration and impulse shaped as damped oscillations. The setup can be deployed in high voltage environments such as substations. The measurement devices used are efficient to record noise samples in the ISM band and our extraction method is able to provide all the information required for channel modeling.

During the measurement campaign, correlation between the power of impulsive noise in terms of amplitude and repetition rate with the voltage of the equipment can be noted. The higher the voltage is, the more impulsive the noise gets. Wireless communications might be more disturbed by substations, especially for the higher voltage, such as 735 kV.

The proposed impulse detection method is very useful for the study of impulsive noise. Indeed, information concerning the impulses such as the duration, the amplitude, the repetition rate and the power spectrum can be extracted. The impulse detection method allows to calculate representative parameters for two-state Markov chain and Middleton class-A models, which can be used when measurement are not available.

The measurement campaigns give a lot of information concerning impulsive noise in substation. Two noise models that can replicate the characteristics of the noise with more accuracy than other models in literature are introduced in next chapters.

Chapter 4
A Physical Model of EMI Induced by a Partial Discharge Source

4.1 Introduction

In last chapters, an overview of electromagnetic interference induced by partial discharges and their impact on wireless communication systems has been discussed. Measurement campaigns in substations have been conducted. These sources can take place in high-voltage equipment like transformers, powerlines, bushing bars, etc. In Chap. 3, these impulsive signals from several partial discharge sources have been characterized using first- and second-order statistics. These allow for the analysis of these typical source in substation. Having an accurate model these EMI sources may be helpful for performance analyses of wireless communication systems in this harsh and hostile environment.

The primary interest in this chapter is to formalize a coherent, detailed, and validated model that links the discharge process to the induced far-field wave propagation. To our knowledge, this is the first complete and coherent approach model that links physical characteristics of high-voltage installations to the induced radio-interference spectrum. The majority of the discharge processes may be caused by partial discharges. This behaviour is distinct from other classical models in the literature. Indeed, in substation environments, the applied voltage is generally AC. As a result, occurrences of discharge events are cyclostationary processes which vary according to the cycle of the AC voltage. Moreover, the emitted radiations occupy a wide range of frequencies. The waveforms received through antennas are transient damped oscillations whose their high amplitude far exceeds the background noise as observed Chap. 3.

In this chapter, we propose the first model of these impulsive electromagnetic interferences whose impulse events are cyclostationary processes and waveforms are transient impulses. We adopt a physical approach to simulate partial discharges in which charge and current densities of interference sources are derived. The proposed model is inspired by models described in [68, 69, 128] in which funda-

© Springer International Publishing AG, part of Springer Nature 2019
B. L. Agba et al., *Wireless Communications for Power Substations:*
RF Characterization and Modeling, Wireless Networks,
https://doi.org/10.1007/978-3-319-91328-5_4

mental aspects of the discharge process have been taken into account. Then, the electromagnetic radiations from charge and current densities can be derived by using an electric dipole approach of interference sources.

The chapter is structured as follows: in Sects. 4.2 and 4.3, we briefly describe the physical aspect of the partial discharges (PD) process that takes place along an insulation surface. Then, in Sect. 4.4, we model an electromagnetic (EM) radiation emitted by a partial discharge source as an electrical dipole in which the interfering source's current of charge density radiates electromagnetic waves. A theoretical formalization of these EM radiations is defined by using retarded potential equations. Inspired by previous works [21, 45, 124], we have defined characterization metrics to compare measurements and the simulation results to validate the model in Sect. 4.5. Measurements are made in the laboratory with a stator bar. Radiations emitted by partial discharge sources are characterized by the phase-resolved partial discharge (PRPD), statistical distributions of power density, inter-arrival time and occurrences. PD sources can also be characterized by power-spectral density (PSD) and impulsive waveforms. Results are discussed in Sect. 4.6 by comparing simulation results and measurements with PD generated by a stator bar at 16 kV rms.

4.2 The Partial Discharge Phenomenon

As described in Chap. 2, a partial discharge is an electrical discharge which partially bridges the insulation between conductors. It is caused by imperfections within or along the insulation surface. These defects could be gaseous inclusions containing voids, cracks in solid materials or bubbles in liquids. From a physical point of view, PD is related to partial electrical breakdown phenomena. When an electric field applied to a dielectric is sufficiently intense, an ionization process occurs that leads to discharge. In addition to a high field stress, an initiatory electron is required to discharge. Hence, a discharge occurs randomly in time, and has a time lag. In [125], a more thorough treatment of the discharge process is presented. The physical phenomenon and its fundamental mechanism are presented in detailed in [49, 50, 53, 68, 69, 128].

A PD is characterized by a sudden drop in the local electric field due to the flow of electric charges, causing a short current-impulse discharge. Radiations emitted by partial discharge sources are random processes in terms of amplitude, the inter-arrival time between two successive impulses and time occurrence. The random character of the process is explained by complex physical mechanisms such as the presence of ionizing radiation, fluctuations in gas density or gas decomposition, etc. as mentioned in [54, 55]. In AC voltages, impulsive events are cyclostationary and occur at every half-cycle of the applied voltage. An example of a basic PD mechanism is shown in Fig. 4.1. When the local electric field $E_0(t)$ due to the applied voltage between electrodes, is sufficiently high to reach an exceeded value defined by E_{inc}, a PD occurs. The impulse train is driven by an AC voltage in this example. Δt_i is the inter-arrival time of subsequent PD events.

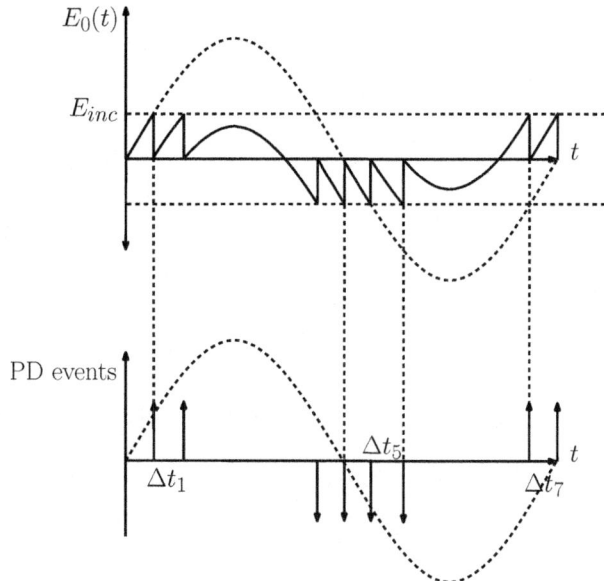

Fig. 4.1 Typical behaviour of discharge process under AC voltage stress

4.3   The Physical Model of Partial Discharge Source

Two physical parameters are required to initiate a discharge: the electric field stress amplitude and the critical value that leads to the ionization process. In this section, we describe a physical model of the discharge process using models of the electric field stress and the discharge occurrence in the ambient air.

4.3.1   Electric Field Stress

The amplitude of the electric field determines the amplitudes and rate of a PD. Depending on the arrangement of electrodes and the presence of the space charge, the electric field distribution can be strongly non-uniform. PDs in substations are formed by non-uniform field distributions due to asperities such as air cavities on insulation surface that have been degraded by PD activities and high stress fields. The electric field along the insulation surface can be determined by an equivalent circuit a telegrapher's equations and illustrated in Fig. 4.2. The presence of space charge is not considered.

Without space charges, the dielectric is represented by its conductivity at the interface between the dielectric surface and air. This is modeled by a resistance R_s per unit length. Its permittivity is modeled by a capacitance C_o per unit length. The term ℓ denotes the width of the air gap.

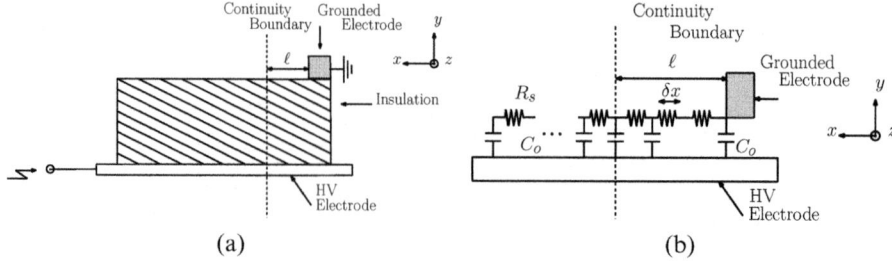

Fig. 4.2 Dielectric between a pair of HV electrodes and its equivalent circuit. (**a**) Surface discharge generation. (**b**) Equivalent circuit

In order to derived the electric field that is tangential to the insulation surface $E_\parallel(x, t)$ at the x coordinate, we first consider the potential on the surface of the bar end denoted as $U(x, t)$ and the electrical current flowing along the surface denoted as $I(x, t)$. These can be defined by the following equations:

$$\frac{\partial U(x, t)}{\partial x} = -R_s I(x, t) \tag{4.1}$$

$$\frac{\partial I(x, t)}{\partial x} = -C_o \frac{\partial U(x, t)}{\partial t} \tag{4.2}$$

Assuming that the potential at the dielectric surface U^* is noted as a phasor such that:

$$U^* = U_0 e^{j\omega t} \tag{4.3}$$

where U^* is a periodic harmonic potential $\in \mathbb{C}$, and U_0 is the magnitude of the AC voltage. We introducing boundary conditions, such that $U(0) = 0$ and $U(\ell) = U_0$ [129, 130]. In a given direction, x, the tangential electric field on the insulation $E_\parallel(x, t)$ is expressed as:

$$E_\parallel(x, t) = -\frac{\partial U^*(x, t)}{\partial x}$$
$$= -\alpha U_0 \frac{\cosh[(\ell - x)\alpha]}{\sinh[\ell\alpha]} \tag{4.4}$$

with $\alpha = \sqrt{j\omega R_s C_o}$. If there are regions in which the electric field exceeds a critical value, an ionization process takes place and PD activity is observed. Furthermore, for a fixed value of C_o, when $R_s \rightarrow 1/C_o$, the field distribution becomes uniform in the cavity and the local field becomes $E \simeq U_0/\ell$. We assume the applied electric field in the air cavity $\mathbf{E}_0(\mathbf{r}, t)$ where $\mathbf{r} = (x, y, z) \in \mathbb{R}^3$ is only determined by the tangential electric field.

4.3.2   Discharge Process

The ionization process of air occurs when the amplitude of the electric field is higher than the critical value. Electrons are released from the surface or at the cathode. The deposited charges on the dielectric surface reduce the local field. As a result, the contribution of PD events to the electric field is given by:

$$\mathbf{E}_i(\mathbf{r}, t) = \mathbf{E}_0(\mathbf{r}, t) - \mathbf{E}_{pd}(\mathbf{r}, t) \tag{4.5}$$

where $\mathbf{E}_{pd}(\mathbf{r}, t)$ is the reduced local field related to the discharge process. In Eq. (4.5), PD activity is approximated by:

$$\mathbf{E}_{pd}(\mathbf{r}, t) = \sum_{n=0}^{\infty} E_{q_n} \delta_{pd}(\mathbf{r}, t - \Delta t_n) \tag{4.6}$$

where $\delta_{pd}(\mathbf{r}, t)$ is a space-time function related to the electric field reduction during the discharge process. E_{q_n} is the contribution of the nth PD, and Δt_n is the inter-arrival time between two consecutive PD events. The critical value depends on a number of factors, such as discharge type, the presence of a cavity or gap, the nature of gas or pressure temperature [70]. The critical value of a streamer discharge in a void can be used to initiate a discharge, although PD does not take place in a spherical void, in keeping with related research [68–70]. The threshold value is a suitable approximation of the PD inception field. This is expressed as:

$$E_{inc} = (E/P)_{cr} P \left[1 + \frac{B}{\sqrt{2P\ell}} \right] \tag{4.7}$$

where $(E/P)_{cr} = 25.2$ VPa^{-1} m^{-1} and $B = 8.6$ m$^{1/2}$ are constant values for the air [69]. P is the atmospheric pressure and ℓ is the air gap width. Thus, PD occurs whenever the local field reaches E_{inc}. Then, during the discharge process, the local field is reduced until its residual field E_{res} is reached. This value is approximately proportional to the critical field [68, 131], such that:

$$E_{res} = \gamma (E/P)_{cr} P \tag{4.8}$$

where γ is the dimensionless factor. Different values of γ can be assumed at different polarities [68]. Under these conditions, by knowing E_{inc} and E_{res}, the local field reduction due to PD activity is expressed as:

$$E_q = E_{inc} - E_{res} \tag{4.9}$$

A high field stress is a necessary but not a sufficient condition to initiate a discharge. The presence of an initiatory electron is required for a PD event. These electrons are produced from surfaces. They are derived from a combination of physical processes such as field emission from cathodic conductors, detrapping of electrons from traps at the insulator surface, collisions between ions, or photo-ionization processes [68]. The surface emission can be modeled by using the Richardson-Schottky formula [68, 70] to model PD activity at insulation surfaces in the following equation:

$$N_e = \frac{A}{e} S_m \exp\left[-\frac{\Phi - \sqrt{eE/(4\pi\varepsilon_0)}}{k_B T} \right] \tag{4.10}$$

where N_e is the emitted electron rate from the surface of area A. e is the elementary charge, Φ is an effective work function, E the electric field at the emitting surface, k_B the Boltzmann constant and T the temperature. S_m is the material surface state. Depending on the insulation part or the conductive part, S_m can be written as:

$$S_{m_c} = C_{th} T^2 \tag{4.11}$$

$$S_{m_i} = v_0 \frac{q}{A} \tag{4.12}$$

where S_{m_c} is the characterization of the conducting part where $C_{th} = 1.2 \times 10^6 \, \text{A m}^{-2} \, \text{K}^{-2}$ is a universal constant. S_{m_i} is the characterization of the insulation part where $v_0 \simeq 10^{14} \, \text{s}^{-1}$ is the fundamental phonon frequency and q is the deposited charge on the insulation by previous PD events [68, 70]. The availability of initiatory electrons on the surface determines the average delay between the time when the field reaches the critical value E_{inc} and the excess value required to initiate a discharge. Thus, PD occurrence can be delayed due to these electrons emitted from the surface. The average time delay is expressed as:

$$\tau_e = \frac{1}{N_e} \tag{4.13}$$

The contribution of the surface emission gives stochastic properties of PD events. The calculation of this time delay was made by [132]. It determines the probability that a discharge may occur after the time when the field strength exceeds its critical value. This effect is explained by the presence of free electrons which start an electron avalanche process. Statistical properties, such as amplitude and time occurrence distributions of the PD activity, are determined by these electrons. The number of the released electrons is assumed to be a random value following the Poisson distribution law [55, 68]. Therefore, the probability of a PD occurrence is given by:

$$p_{pd} = 1 - p_{\tau_e} \tag{4.14}$$

where p_{pd} is the probability of a PD occurrence and p_{τ_e} the probability of delayed electron emission from the surface.

4.3.3   Current and Charge Density

When a PD occurs, charged particles move from cathode to anode via a conductive spark streamer channel or electrical arc. During the discharge process, the electrical arc is seen as a conductor. Under this condition, the current density $\mathbf{J}(\mathbf{r}, t)$ flows uniformly in the conductor. We assume the ionized air to be a plasma [133]. We link the current density with the electric field reduction and the plasma conductivity σ by:

$$\mathbf{J}(\mathbf{r}, t) = \sigma \mathbf{E}_{pd}(\mathbf{r}, t) \tag{4.15}$$

The plasma conductivity is written as:

$$\sigma = \frac{Ne^2}{m_e v_c} \tag{4.16}$$

where N is the density of electrons, e is the elementary charge, m_e is the electron mass and $1/v_c$ is the mean scattering time between subsequent charge collisions. The charge density induced by a discharge can be written using the expression of the electric field induced by the discharge process in keeping with Gauss's law, such that:

$$\rho(\mathbf{r}, t) = \varepsilon_0 \nabla \cdot \mathbf{E}_{pd}(\mathbf{r}, t) \tag{4.17}$$

where $\nabla\cdot$ is the divergence operator and ε_0 the vacuum permittivity.

4.4   The Electromagnetic Radiation of the Interference Source Induced by Partial Discharge

Due to the local ionization process, the ionized air becomes a conductor. Hence, charges and currents are produced from the interference source. From the Lorenz gauge equation, electromagnetic radiations can be derived from current and charge densities. In this section, we assume a PD source to be an electric dipole. In the far field region, EM radiations and power density are derived from the retarded potentials equations. Inspired by models in [37, 134], as well as many experimental results in [18, 135, 136], we propose a coherent approach to model electromagnetic impulsive transient noise interferences induced by corona discharges on high-voltage installations.

4.4.1   Electric Dipole formulation

Due to the movement of charges, we assume that the ionization process leads to a conducting channel in which currents and charges generate electromagnetic wave radiations. Therefore, from the local ionization area, there is a magnetic potential vector source and an electric scalar potential source produced by a current density and a charge density, respectively. From an antenna located at a point $\mathbf{r}$ in the three-dimensional space in the far field region, potentials are written by solving Lorenz gauge condition equations defined by:

$$\nabla \cdot A(\mathbf{r}, t) + \varepsilon \mu \frac{\partial V(\mathbf{r}, t)}{\partial t} = 0 \qquad (4.18)$$

where $A(\mathbf{r}, t)$ and $V(\mathbf{r}, t)$ are respectively the magnetic vector potential and the electric potential of the discharge source. By assuming the harmonic time dependence, potentials $\mathbf{A}(\mathbf{r}, t)$ and $V(\mathbf{r}, t)$ are written from the solution to the Eq. (4.18) respectively as:

$$\mathbf{A}(\mathbf{r}, t) = \frac{\mu}{4\pi} \int_{\mathbf{v}'} \mathbf{J}(\mathbf{r}', t - t') \frac{e^{-j\omega\sqrt{\varepsilon\mu}|\mathbf{r} - \mathbf{r}'|}}{|\mathbf{r} - \mathbf{r}'|} d\mathbf{v}' \qquad (4.19)$$

$$V(\mathbf{r}, t) = \frac{1}{4\pi\varepsilon} \int_{\mathbf{v}'} \rho(\mathbf{r}', t - t') \frac{e^{-j\omega\sqrt{\varepsilon\mu}|\mathbf{r} - \mathbf{r}'|}}{|\mathbf{r} - \mathbf{r}'|} d\mathbf{v}' \qquad (4.20)$$

where $\mathbf{J}(\mathbf{r}, t)$ is the current density and $\rho(\mathbf{r}, t)$ the charge density produced by the interference source, and $\mathbf{v}'$ is the volume in which the induced charges are confined. The delayed time propagation from the interference source to an antenna respectively positioned at $\mathbf{r}'$ and $\mathbf{r}$ is represented by $t' = |\mathbf{r}' - \mathbf{r}|/c$, where c is the speed of light in vacuum. These retarded potentials equations can be seen as spatial convolution products by Green's functions.

It is worth remembering that during the discharge process, an electrical arc is a conducting channel where charge and current densities generate electromagnetic waves. Consequently, it is seen as a wire conductor when its length ζ is lower than the wavelength λ. Based on Eq. (4.6), the expression of the current impulse $I_{pd}(t)$ produced by the interference source is written as:

$$I_{pd}(t) = \sum_{n=0}^{\infty} \sigma E_{q_n} \delta_{pd}(t - \Delta t_n) \qquad (4.21)$$

The integration in Eq. (4.19) allows us to link the expression of the current $I_{pd}(t)$ and the length of the wire conductor ζ by $I_{pd}(t)\zeta$. Now, the magnetic- and electric-radiated fields can be determined by using:

$$\mathbf{H}_r(\mathbf{r}, t) = \frac{1}{\mu} \nabla \times \mathbf{A}(\mathbf{r}, t) \qquad (4.22)$$

$$\mathbf{E}_r(\mathbf{r}, t) = -\nabla V(\mathbf{r}, t) - \frac{\partial \mathbf{A}(\mathbf{r}, t)}{\partial t} \qquad (4.23)$$

We place an antenna in the far field region where $|r| \gg \lambda$. The radiated fields by PD activity in spherical coordinates $\mathbf{e}_{sph} = (\mathbf{e}_r, \mathbf{e}_\theta, \mathbf{e}_\varphi)$ are:

$$\mathbf{H}_r(r, \theta, t) = \frac{j\beta_0 I_{pd}(t)\zeta}{4\pi r} \sin\theta e^{-j\beta_0 r} \cdot \mathbf{e}_\varphi \tag{4.24}$$

$$\mathbf{E}_r(r, \theta, t) = \sqrt{\frac{\mu_0}{\varepsilon_0}} \frac{j\beta_0 I_{pd}(t)\zeta}{4\pi r} \sin\theta e^{-j\beta_0 r} \cdot \mathbf{e}_\theta \tag{4.25}$$

where $\beta_0 = \omega\sqrt{\varepsilon_0\mu_0}$.

4.4.2   Power Radiation of the Interference Source Received at the Antenna

By assuming isotropic radiations in the far field region, the Poynting theorem allows for the expression of the instantaneous power radiation emitted by the interference source as:

$$
\begin{aligned}
P_{pd}(t) &= \int_S \mathbf{S}(r, \theta, t)dS \\
&= \int_0^{2\pi} d\varphi \int_0^\pi Z_0 |I_{pd}(t)|^2 \left(\frac{\beta_0\zeta}{4\pi r}\right)^2 \sin^3\theta r^2 d\theta \\
&= 2\pi Z_0 \frac{|I_{pd}(t)|^2}{3} \left(\frac{\zeta}{\lambda}\right)^2
\end{aligned}
\tag{4.26}
$$

where $\mathbf{S}(r, \theta, t) = \mathbf{E}_r(r, \theta, t) \times \mathbf{H}_r^*(r, \theta, t)$ is the Poynting vector of PD radiation and $Z_0 = \sqrt{\mu_0/\varepsilon_0} = 120\pi \ \Omega$. Since impulsive noise radiations are characterized by a fast transient component, each impulse can be determined by its power density $P_d(t_n)$ such that:

$$P_d(t_n) = \frac{1}{4\pi r^2} \int_{t_n - t_w/2}^{t_n + t_w/2} P_{pd}(t)dt \tag{4.27}$$

where r is the distance between the interference source and the antenna, t_w is the duration of each impulse and t_n is the time occurrence. If the observation point is an electric dipole such as an antenna, the radiative power received depends on its effective surface or effective aperture A_e, defined by:

$$P_r(\theta, \varphi) = A_e(\theta, \varphi)P_d \tag{4.28}$$

where θ and φ are angular coordinates of the beam.

4.4.3   Modeling Impulsive Waveforms and PSD

Impulsive waveforms emitted by partial discharges can be modeled numerically by using a linear time-invariant (LTI) filter, defined in the z-domain as:

$$U(z) = H_m(z)\epsilon(z) \qquad (4.29)$$

where $\epsilon(z)$ is the input of the LTI defined as a Dirac impulse in the time-domain, $U(z)$ is the resulting impulsive waveform and $H_m(z)$ is the LTI filter. The latter includes both the frequency response of the measurement setup and the discharge. We have interpolated the frequency response of the measurement setup according to data sheets. $H_m(z)$ is approximated by a digital resonator written as:

$$H_m(z) = G \frac{1 - z^{-2}}{1 - (2b_0 \cos \omega_0) z^{-1} + b_0^2 z^{-2}} H_{ms}(z) \qquad (4.30)$$

where $H_{ms}(z)$ is given by the frequency response of the measurement setup, ω_0 is the resonant frequency, b_0 gives the bandwidth and G is the normalization gain of the filter given by:

$$G = \frac{(1 - b_0)\sqrt{1 + b_0^2 - 2b_0 \cos 2\omega_0}}{\sqrt{2(1 - \cos 2\omega_0)}} \qquad (4.31)$$

Parameters of $H_m(z)$ will be set in Sect. 4.6. The model will be compared to experimental results in terms of average power spectral density (PSD) and waveforms in the discussion.

4.4.4   Brief Summary of Interference Induced by Discharge Source

The interference source can be summarized by the following statements:

- a high electric field stress and an initiatory first electron are required to initiate discharges;
- the availability of these delayed electron emissions affects the stochastic properties of impulses in terms of power radiation, time occurrence and inter-arrival time;
- a PD activity reduces the local electric field due to deposited charges;
- discharge events follow the cyclostationary process when an AC voltage is applied;
- during the discharge process, charge and the current densities are produced by an electrical arc;

- the electromagnetic radiations are derived from charge and the current densities during the discharge process. An antenna can receive these radiations as an impulsive transient noise.

An example of a computation of a PD process occurring at every rise of the AC voltage is illustrated in Algorithm 1, with the presence of a first electron to initiate discharges. The following notations are introduced for the implementation of partial discharges :

- l is the number of the cycle of the AC voltage;
- t_{inc+} and t_{inc-} are the time of the first PD occurrence during the first half-cycle and the second, respectively;
- t_n is the time occurrence of the nth impulse;
- $E_0(t)$ is the local field applied to the electrodes, which is a AC voltage;
- $E_q = E_{inc} - E_{res}$ is the local field reduction due to the PD process;
- $\Delta E = E_{inc} + E_{res}$ corresponds to the contribution of the electric field between the last PD event and the first impulse at the last half-cycle;
- $\mathbf{S}(r, \theta, t)$ is the Poynting vector;
- $P_{pd}(t)$ is the power density of impulsive radiations emitted by a partial discharge source.

Algorithm 1 Computation of PD process sequence during the first half-cycle

---

Ensure: $\exists t_{n-1} \in \left[t_{inc-} ; (2l - 1)\pi + \frac{\pi}{2} \right]$, a PD impulse exists for $l > 0$;
  if $l > 0$ **then**
    if $\exists t_n \in \left[t_{n-1} ; 2l\pi + \frac{\pi}{2} \right]$, $E_0(t_n) - E_0(t_{n-1}) = \Delta E$ **then**
      $P_{pd}(t_n) \leftarrow \int_S \mathbf{S}(r, \theta, t_n) dS$;
    end if
    $n \leftarrow n + 1$
  end if
  while $t_{n-1} < t \leq 2l\pi + \frac{\pi}{2}$ **do**
    if $t_n, \in t$, $E_0(t_n) - E_0(t_{n-1}) = E_q$ **then**
      $P_{pd}(t_n) \leftarrow \int_S \mathbf{S}(r, \theta, t_n) dS$;
      $n \leftarrow n + 1$;
    else
      $P_{pd}(t) \leftarrow 0$;
    end if
  end while

4.5   Experimental Characterization Process of the Interference Source

4.5.1   Definition of Characterization Metrics

Based on Chap. 3, the proposed model can be validated according to the characterization of PD sources as follows:

- power density of impulsive noise radiations P_d distribution (mW/m^2);
- PD occurrence t_n distribution (s);
- inter-arrival time Δt distribution (s).

These first order statistics allow for the analysis of PD occurrence and its power density distribution per unit time or phase. For an isotropic antenna, the measured signal in voltage $V_m(t)$ is converted into electric field strength $x_m(t)$ defined as in Eq. (3.6).

In practice, the measured signal contains impulsive noise emitted by discharges and the overall background noise. The impulse detector is used for an experimental characterization of PD as presented in Chap. 3. First, impulsive noise emitted by partial discharge sources is extracted from the overall noise via a denoising process. Then, the power density of each impulse is calculated by using a short-time Fourier analysis. PD processes are characterized in terms of statistical distributions and also in terms of power spectral density (dBW/m^2/Hz).

4.5.2   Denoising Process

The power spectral density of the radiations that are received from partial discharge sources can be estimated properly by using a denoising process that extract impulses from the overall noise. The denoising process is described in Chap. 3. In this experimentation, Daubechies wavelets are used with eight vanishing moments and for 30 levels of decomposition to yield a suitable estimation of impulsive transient noise waveforms [124].

4.5.3   Short-Time Analysis Process

The measurement data contain a train of impulses located randomly in time. PD can be detected by using a short-time analysis of the spectrogram of the waveform. The procedure is presented in Chap. 3. To analyse each impulse in a given impulse train, we use a wide band spectrogram with a Hamming window of $t_w = 6.4$ ns length. The length of the FFT is $N_{\text{fft}} = 2048$ and 50% overlap is used. Under this condition, the measured power of the ambient noise without discharges is about

-53.20 dBW/m^2. The power density of each impulse is calculated in the bandwidth range of 0.2–5 GHz. $P_d(t_n)$ value is located at each signal peak above -47 dBW/m^2 to ensure the detection of significant impulsive noise emitted from partial discharge sources.

4.6 Experimental Validation

In this section, the experimental validation is based on the comparison of measurement campaign and the simulation results. The following subsections describe the measurement setup including the PD sources produced by a stator bar. Then, the simulation parameters are adapted according to measurement conditions. Note that the measurement setup follows Chap. 3.

4.6.1 Brief Description of Measurement Setup

4.6.1.1 The Measurement Setup

In this chapter, the observation time is 20 ms at a 10 Gs/s sample rate (200 M samples). The oscilloscope is synchronized to the HV voltage generator. Waveforms are captured periodically at the beginning of each cycle of the AC voltage. Using this configuration, 30 waveforms are captured. The antenna is positioned at $d = 3$ m from the stator bar in the far field region.

4.6.1.2 PD Sources from Stator Bar

A stator bar is used to generate PDs along the insulation surface. A main PD site is located on the insulation surface as illustrated in Fig. 4.3. In the middle of the bar, the semi-conducting coating has been removed to expose the epoxy-mica insulation. An air gap between insulation and grounded electrode, including the semi-conducting coating, can be seen. The Dimensions and electrical properties of the dielectric and air cavities used are summarized in Table 4.1 and Fig. 4.4. The conductivity of the insulation surface is attributed to the presence of epoxy-mica. The conductivity of the epoxy-mica has been measured in [56]. It can be found that σ_i can be $\sim 10^{-16}$ S/m for unaged dielectric, and $\sim 10^{-9}$ S/m for aged dielectric. The air cavity is represented by the thickness of the semi-conducting coating on the bar. Thus, when the electric stress is sufficiently high, PD activities can take place at the end of the semi-conducting coating. A HV generator is applied to the stator bar. Surface discharges are predominant in this experimentation. PDs are generated by an AC high voltage of 16 kV rms at 60 Hz.

Fig. 4.3 Stator bar and measurement configuration

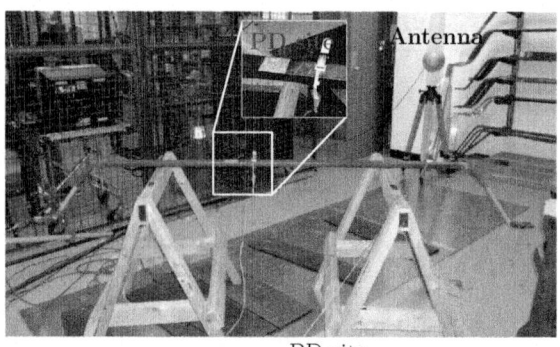

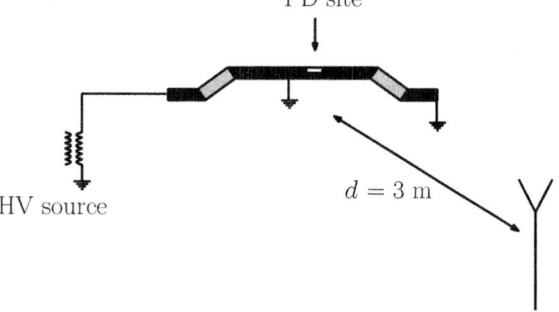

Table 4.1 Dimensions and electrical properties of PD process

| Model | Dimensions | | | Relative permittivity | Conductivity |
| | Height | Width | Depth | | σ_i |
	(mm)	(mm)	(mm)	ε_r	(S/m)
Air cavity	0.25	2.5	2	1	0
Dielectric	3.6	55	58	4	2.8×10^{-9}

4.6.2   Simulation Setup

Simulation parameters are defined by the following steps:

4.6.2.1   Calculation of the Electric Field Along the Surface

The electric field inside air cavities is calculated from the equivalent circuit where r_s and c_0 are the resistance of the insulation surface and the capacitance of the grounded wall, respectively. Depending on the orientation of air cavities, ℓ is in the x or z coordinate. These components are written as:

Fig. 4.4 The surface of a
stator bar with modeled air
cavities. (**a**) PD site with
modeled air gap cavities. (**b**)
Dimensions of stator bar with
an air cavity

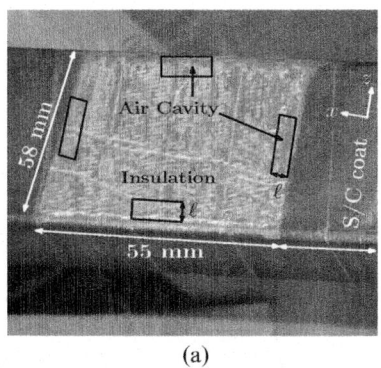

(a)

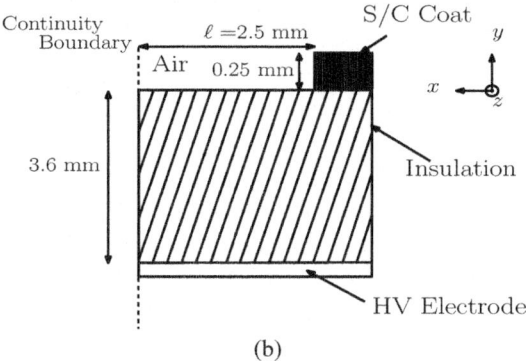

(b)

$$r_s = \frac{1}{\sigma_i A_r}$$

$$c_0 = \frac{\varepsilon_0 \varepsilon_r A_c}{\eta \delta x}$$

(4.32)

with $\delta x = \delta z = 0.5$ mm, calculating the cross sectional area of the resistance A_r is the product of the depth of the dielectric (58 mm) and its height (3.6 mm), the calculated resistance per unit length is $r_s = 1.72$ TΩ/m. Calculating the overlapping surface area of the plates A_c as the product of δx and the depth of the dielectric (58 mm). Since η is the height of the dielectric (3.6 mm), the capacitance of the insulation is $c_0 = 0.604$ nF/m. By applying an AC voltage at 16 kV rms, the tangential electric field is calculated from Eq. (4.4). Since the tangential electric field is complex, we use the absolute value to determine its amplitude.

4.6.2.2 Discharge Process in Air Cavity Parameters

These parameters have been set in keeping with several related works [68, 69, 128]. PD activities take place on either side of the air gap between the end of the semi-

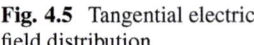

Fig. 4.5 Tangential electric field distribution

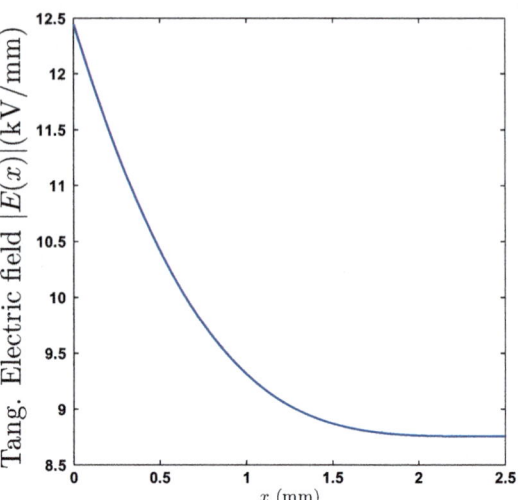

conducting coating and the insulation surface. In Fig. 4.4, four air cavities are considered. Although many PD sources can take place randomly in these air cavities, four PD sources are assumed for each. PD sources are located in the middle of the high-stress region. At 16 kV rms voltage, the electric field amplitude at $x = 0$ is 12.45 kV/mm as depicted in Fig. 4.5. From Eq. (4.7), under normal temperature and pressure, the critical value to initiate a discharge is $E_{inc} = 3.96$ kV/mm. Since the calculated electric field amplitude is higher than the E_{inc}, PD activity is observed.

We characterize PD amplitudes by the local field reduction, as defined in Eqs. (4.6) and (4.9). The reduction depends on the deposited charges in the air cavity. Previous discharges and other discharges in the vicinity of the air cavity can also affect this local field reduction [55, 70]. The latter determine the behaviour of the inter-arrival time and also the amplitude of the field reduction. We assume that E_q is a random value following an exponential distribution, where the average value is:

$$E_{q^+} = E_{inc} - \gamma_+(E/P)_{cr}\, P$$
$$E_{q^-} = E_{inc} - \gamma_-(E/P)_{cr}\, P \qquad (4.33)$$

where E_{q^+} and E_{q^-} are the local field reduction value of the conductive part and the insulation part, respectively, where $\gamma_+ = 0.01$ and $\gamma_+ = 1.2$.

The release of free electrons from surfaces is given by Eq. (4.10). We define the conductive part and the insulation part by their work functions Φ_c and Φ_i, respectively. Although emission surface parameters could not be measured directly, they have been adapted to reproduce measurements. We use $\Phi_c = 1.20\ eV$, $\tau_{e_c} \simeq 36\,\mu s$ and $\Phi_i = 1.38\ eV$, $\tau_{e_s} \simeq 21\,\mu s$.

4.6.2.3 Stochastic Property of the Emitted Radiations of PD Sources

Due to the reduction of the local field, each air cavity generates charge and current density according to Eqs. (4.15) and (4.17). The retarded potentials equations allow us to determine the induced electromagnetic radiations from partial discharge sources, as defined in Eq. (4.27). The amplitude of the electromagnetic field is derived from the currents and the length of the electric arc, $I_{pd}(t)\zeta$. The amplitude of the current impulse depends on the local field reduction value E_q and the conductivity of the ionized air σ as seen in Eq. (4.21).

The conductivity σ as defined in Eq. (4.16) depends on the collision frequency between species and the density of electrons. We assume that the combination of many PD sources affects the plasma conductivity randomly and independently. Moreover, due to the applied voltage, the discharge process is time-dependent in terms of amplitude and occurrence. Therefore, we express the plasma conductivity σ as:

$$\sigma^+(\sigma_0^+, t) = \sigma_0^+ f^+(t)$$
$$\sigma^-(\sigma_0^-, t) = \sigma_0^- f^-(t)$$

(4.34)

where σ_0^+ and σ_0^- are the average random values. $f^+(t)$ and $f^-(t)$ are deterministic time-dependent functions related to the influence of the electric field stress on charge density and the mean scattering time. These functions determine the time occurrence distribution of PD events. It may be challenging to obtain parameters of plasma conductivity. Inspired by relevant models [23, 37, 45], we have adapted these parameters with observations from the measurement campaign. Thus, we choose that the quantity $\zeta\sigma$ can be considered as a random value according to Weibull distributions. Parameters of the distribution are estimated from empirical data obtained from the measurement campaign. In addition, we have decided that time-dependent functions $f^+(t)$ and $f^-(t)$ are Gaussian functions. The antenna is located 3 m from the source, and 30 cycles of the 60 Hz are considered.

4.6.3 Simulation-Measurement Comparison

4.6.3.1 PRPD Comparison

In this chapter, the discussion is restricted to the comparison at 16 kV rms. The power density of these impulsive radiations and their occurrences can be compared in terms of the measurement and simulation results by PRPD illustrated in Fig. 4.6. According to measurements, PD events occur at every half cycle of the AC voltage. These results show the cyclostationary process of PD sources generated by an AC voltage. The impulse rate and the power density in the negative part is higher compared to what occurs in the positive part. This is due to the number of deposited

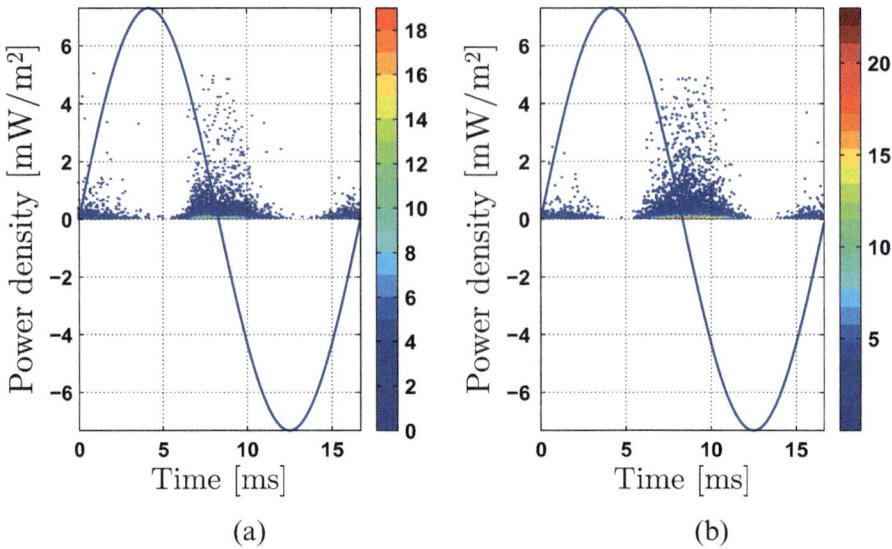

Fig. 4.6 Phase resolved partial discharge 16 kV rms. (**a**) Measurement. (**b**) Simulation

charges on the insulation surface. They can be explained by a low value of the electric field reduction on the cavity $E_q^- < E_q^+$ and a higher value of the air conductivity due to the ionization process.

A high probability of having a PD with low amplitude of power density is observed and the probability of having a high amplitude of power density is lower. This is because the local field of PD activity has yielded a low value of reduction. Moreover, the emission of electrons from surfaces influences the amplitude of the power density and the rate of PD events. When the critical field stress E_{inc} is reached, the presence of the initiatory electrons is not sufficient to discharge. Consequently, a time delay is observed due to the excessive electric field stress-to-discharge process, which causes the power density of impulsive radiations to increase. The obtained PRPD from the simulation results matches with measurement results. Both results indicate that the cyclostationary process of PD events are induced by the applied voltage. The occurrence of discharge events is predominant at the negative polarity compared to the positive one.

4.6.3.2   Statistical Distributions Comparison

The statistical distributions of power density, inter-arrival time and time occurrence for measurements and the simulation at 16 kV rms are presented as an empirical probability density function (PDF) and empirical complementary cumulative distribution function (CCDF) in Figs. 4.7, 4.8 and 4.9. Statistical results shows that PD

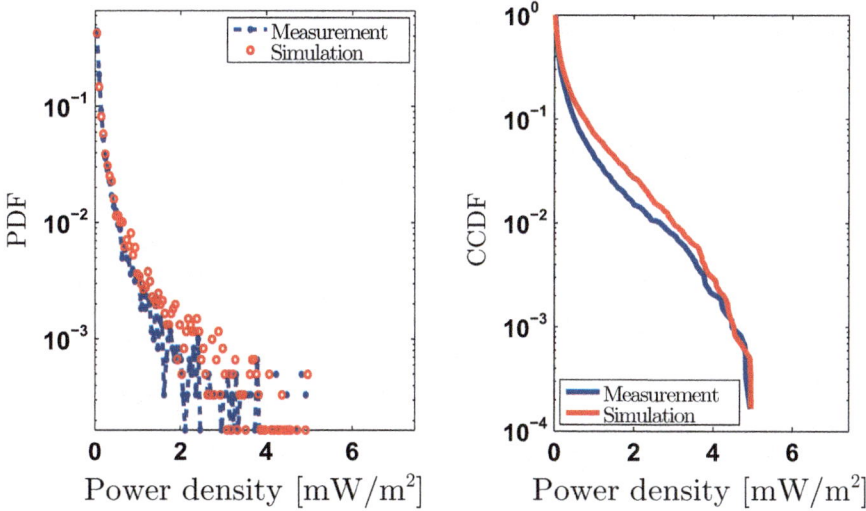

Fig. 4.7 Distribution of power density: measurement vs. simulation

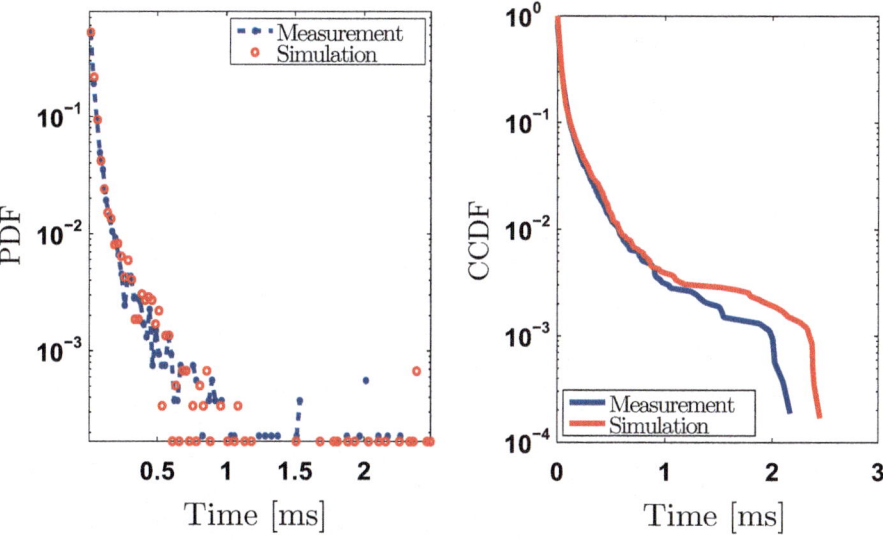

Fig. 4.8 Distribution of inter-arrival time: measurement vs. simulation

events can be described as random processes in terms of power density amplitude, inter-arrival time and time occurrence.

Stochastic properties of PD events are explained by physical interactions between charged particles, such when these particles recombine in order to reach their equilibrium state. These interactions can take place because of previous discharges

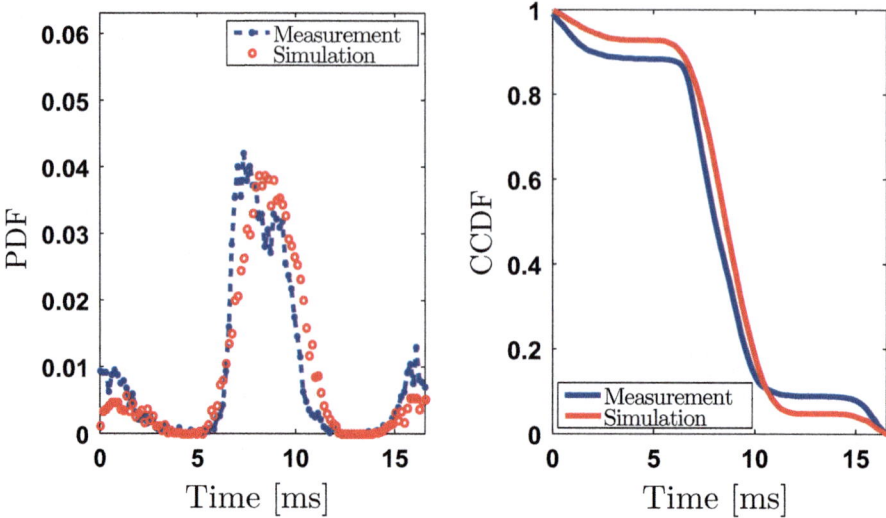

Fig. 4.9 Distribution of time occurrence: measurement vs. simulation

Table 4.2 The goodness-of-fit: measurement vs. simulation

Test statistics	Power density	Inter-arrival time	Time occurrence
D_{KL}	0.1873	0.1706	0.2153
D_{KS}	0.0687	0.0077	0.127

or as a result of the PD activities of other sources. The excessive electric field stress of the applied voltage increases the probability of releasing free electrons and charged particles by collision into the air. As a result, the air becomes ionized.

Figure 4.7 shows power density distributions of measurement and simulation. The probability of a high amplitude of power density decreases exponentially. The inter-arrival time distributions follow exponential distributions in measurement and simulation results (see Fig. 4.8). Figure 4.9 shows time occurrence distributions. PD events are observed as the applied voltage rises and falls. They follow the cyclostationary process induced by the frequency of the applied voltage. The probability of observing a discharge at the negative polarity is higher than the positive. Time occurrence PDF describes two Gaussian distributions. We compare measurement and simulation results by plotting the empirical CCDF, measuring the Kullback-Liebler (KL) divergence of the empirical PDF, and conducting a Kolmogorov-Smirnov (KS) test at 5% significance level. It is observed that simulation results fit measurement results. In Table 4.2, small KL-divergence values show a small divergence between PDFs obtained by simulation and measurement. Using the test statistic values of the KS test (D_{KS}) in this table, the obtained p-values are 0.61, 0.72 and 0.37 for power density, inter-arrival time and time occurrence respectively. As a result, the test does not reject the null hypothesis that simulation and measurement results come from the same distributions at 95%. Therefore, simulation results are in agreement with experimental results in terms of first-order statistics.

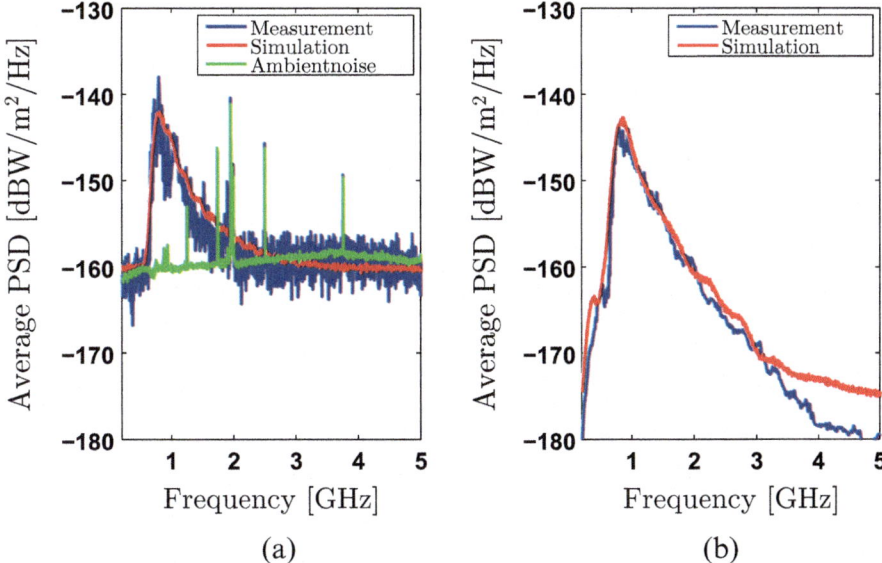

Fig. 4.10 PSD comparison: measurement vs. simulation. (**a**) Raw PSD. (**b**) Denoised PSD

4.6.3.3 PSD and Waveforms of Impulses

Simulation results validate the proposed model in terms of PSD and waveforms. The average PSD of measurement and simulation are presented in Fig. 4.10. The ambient noise includes thermal noise, RF communications at 1.7 and 1.9 GHz and harmonics at 1.25, 2.5 and 3.75 GHz caused by interleaving artefacts and clock feedthrough from the oscilloscope. The PSD of impulses can be extracted from the ambient noise using a denoising process as depicted in Fig. 4.10b. Low wavelet coefficient values have been removed via a hard threshold to extract the impulsive signal. However, the process can significantly modify the PSD of impulses mainly at frequencies at which the PSD of ambient noise is dominant.

PSD and waveforms can be provided by the proposed LTI model in Eq. (4.30). We have considered the ambient noise by adding a Gaussian white noise. Parameters of the LTI filter have been set as follows: $\omega_0 = 0.5027$ (800 MHz) and $b_0 = 0.90$ (bandwidth of 273.2 MHz at -3 dB). By comparing the average PSDs, simulation results match experimentation results (see Fig. 4.10). However, the proposed model cannot perform distortions of waveforms as depicted in Fig. 4.11. These distortions may be caused to multiple reflections of EM waves in the laboratory. Furthermore, in this example, the denoised waveforms and their induced PSDs are presented. Their average energies are approximately equal. It is seen that the measured waveform and the PSD are widely distorted.

The proposed LTI filter models the impulse response of RF-signals from PD activity on average. The filter can be improved by taking into account multiple

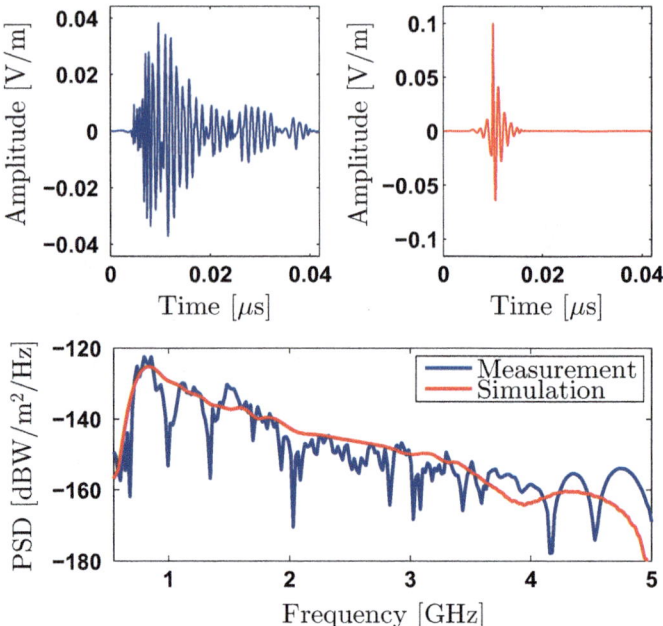

Fig. 4.11 Waveforms and PSD of an impulse

delayed impulses. This can be done by using tapped delay line filters. The impulse response of the propagation channel should be estimated using measurements in the laboratory. Thus, a realistic impulsive RF signals model can be obtained under this condition.

4.7   Conclusion

A physical model of EMIs induced by corona discharges is presented in this chapter. A high stress field can produce an ionization process that leads to discharge. Charges and current densities induce electromagnetic radiations. The interference sources induced by PDs have particular behaviours that differ in terms of occurrences and amplitude from classical models of impulsive EMIs sources [23, 25, 137]. Indeed, the AC voltages produce a cyclostationary behaviour of discharge events. Their amplitudes are induced by the local reduction of the electric field and charged particles during the discharge process.

The physical model allows for a coherent, detailed, and validated approach that links the discharge process to the induced far field wave propagation. As a result, we can link certain physical characteristics of high-voltage installations to the induced radio interference spectrum. We have shown the consistency of simulation results, provided by the proposed model, over experimental results. We have validated the model in terms of amplitude i.e. power density, occurrences, and inter-arrival time of these impulsive radiations. We have also validated the induced spectrum and waveforms of these radiations received at the antenna.

Future work may focus on extending the proposed model to contexts in which there are many discharges with various parameters, such as electric field stress, where the insulation property of dielectrics can reproduce the environment of substations, or wherever else high-voltage installations and their induced discharged sources are randomly distributed in space. Moreover, based on the proposed LTI filter, modeling impulsive waveforms can be significantly improved by taking into account multiple delayed impulses.

Chapter 5
Analysis and Modeling of Wideband RF Signals Induced by PD Using Second-Order Statistics

5.1 Introduction

In several measurement campaigns in substations reported in Chap. 3, wideband impulsive noise waveforms emitted by partial discharges are transient, their samples are correlated and their power spectral densities have approximately a form of $\sim(f - f_0)^{-\gamma}$, where f_0 is the resonant frequency of the RF measurement setup and $\gamma > 0$ an arbitrary exponent. As seen in Chap. 2, one of the most common impulsive noise models used are Middleton Class A and B [24, 25, 91] and the α-stable noise models [77, 138, 139]. Although these models can approximate noise samples with impulsive noise based on first-order statistic, they are far from the observations. Indeed, an impulse is modeled by a single sample. Thus, second-order statistics are, most of the time, inaccurate in the sense that the autocorrelation of an impulsive component yields a Dirac impulse function at zero so that the power spectral density is constant over all frequencies (i.e. not proportional to $f^{-\gamma}$).

Alternatively, a research study reported in [23] has found that transient impulsive noise can be modeled based on a partitioned Markov chain. The model can approximate second-order statistics from the measurements. However, the definition of a suitable number of states in the Markov chain can be challenging and the estimation procedure are complex [83, 140]. On the other hand, impulsive noise can be modeled using Markov-Middleton or Gauss-Markov impulsive noise models as presented in [26, 83], which are used in order to account for the bursty nature of impulsive noises. However, the resulting impulsive waveforms are modeled by a group of uncorrelated samples. Therefore, the transient behaviour of the impulses is not taken into account and the modeled waveforms might be inaccurate. These impulsive noise models are physically limited because their parameters cannot provide any information linking the physical characteristics of HV installations to the induced radio interference spectrum.

© Springer International Publishing AG, part of Springer Nature 2019
B. L. Agba et al., *Wireless Communications for Power Substations: RF Characterization and Modeling*, Wireless Networks,
https://doi.org/10.1007/978-3-319-91328-5_5

A simpler way to model impulsive transient waveforms is to use LTI filters as developed by [141, 142]. Although these filters can capture spectral characteristics of the impulsive radio interference spectrum in substations, the resulting waveform is deterministic, which is not observed in practice. A study of propagation effects of wideband radiation signals from PD activity was investigated by [63]. It has been shown that measured impulsive waveforms are, in general, distorted by the effect of multipath in electromagnetic wave propagation. This is due to the presence of multiple reflectors in substations. Therefore, we may assume that PDs in HV equipment have spectral signatures whose resulting impulsive radio interference spectra are distorted by multipath propagation effects.

5.1.1   Main Contribution and Organization

In this chapter, we propose a novel approach to modeling impulsive noise for wireless channels. By this approach, modeled waveforms stochastically are similar to the measurements in terms of stochastic metrics, in that spectral characteristics of RF signals from PD sources can be captured, and the effect of multipath is taken into account. Compared to many approaches presented in the literature, such a model can reproduce transient effects of impulsive noise, and it fits accurately measurements in terms of second-order statistics. Moreover, a well-established validation procedure can be implemented for the estimation of the spectral characteristics and the selection of a suitable number of parameters in the LTI filter [143–147]. The proposed impulsive noise model allows for the study of the spectrum of radio signals from PD activity and its relationships with physical characteristics of PD sites. This can be used to represent a generic environment of substations for performance analysis of wireless communication networks, as well as for PD detection methods.

The rest of this chapter is organized as follows: in Sect. 5.2, the measurement setup and two major pieces of HV equipment in a 735 kV substation are briefly presented. In Sect. 5.3, a review of measured impulsive noise is presented from this two HV equipment. This shows that an HV equipment has a spectral signature due to the spectral characteristics of the emitted PD. In Sect. 5.4, a generalized model is proposed using an LTI filter approach, by which spectral characteristics of PD sources in HV equipment can be captured. In Sect. 5.5, the adequacy of the approach (the goodness-of-fit) can be measured by analysing the residuals of fitted ARMA models whose distributions are non-Gaussian and whose variances are time-dependent. The impulsiveness of the PD transient waveforms can be reproduced by assuming the conditional heteroskedasticity of the residuals. Finally, in Sect. 5.6, computer simulations are provided to validate the effectiveness of the proposed model vis-à-vis impulsive waveforms measured in substations. The advantages and the limitations of the approach are also discussed.

5.2 Measurement Setup

In the measurement campaigns, impulses are measured employing the measurement setup explained in Chap. 3. An antenna is positioned in the far field region of discharge sources. Measurements are made in a 735 kV outdoor substation, in which EMI produced by two typical HV installations: (a) a power transformer tank (PTT) in which the applied voltage is 735 kV, and (b) a current transformer tank (CTT). The measurement setup described in Chap. 3 has been used. Significant impulsive radiations emitted by discharge sources are detected and captured by a given threshold slightly above the level of the background noise. The sample rate is 10 Gs/s for a given observation time according to the durations of impulses. From experimentations, significant impulses are captured above -33 dBW/m^2 for an observation time of approximately 50 ns in (a) and -42 dBW/m^2 for an observation time of 50 ns in (b).

5.3 Conjectures and Mathematical Formulation of EM Waves

This section provides a review of measured impulsive noises, a discussion of their properties, and a generalization of the observations. Radiations emitted by discharge sources are impulsive, have short durations and exhibit a damped effect which decays over time. We analyse measured waveforms by using a time-frequency representation and second-order statistics, using the autocorrelation function and power spectral density. The analysis may be helpful for the design of a general model of impulsive noise induced by electric arc discharges.

5.3.1 Second-Order Statistics

5.3.1.1 Time-Frequency Analysis

It may be useful to have a time-frequency representation of transient waveforms. Under this condition, we choose the spectrogram representation, which is based on the short-time Fourier transform (STFT) as presented in Chap. 3. In order to have a suitable time-frequency resolution, we compute the spectrogram by considering $w(t)$ as a Hamming function with a window length of 32 samples. The overlap percentage between adjoining sections is 70% of 32 samples and the number of FFT is 512.

5.3.1.2 Autocorrelation Function

Sample autocorrelation is used to estimate the theoretical autocorrelation function
(ACF) as described in Chap. 3. It should be noted that although Gauss-Markov
processes take into account the correlation, they are only one particular case of a
first-order autoregressive process, AR(1). The ACF function of the measurements
suggests that a high order of ARMA processes might be more accurate.

5.3.1.3 Results from the Measurement Campaigns

A typical example of waveforms captured in the measurement campaigns is depicted
in Figs. 5.1 and 5.2. Spectrograms and power spectral densities are also depicted.
The spectrogram shows that the power of transient waveform is strong for a short
duration with a large frequency bandwidth, but decreases over both time and
frequency. Since the noise samples are correlated, the power spectral densities are
non-constant over frequencies. Indeed, the power spectral density (PSD) decays
over high frequencies. As shown in several studies [21, 41, 63, 148], local variations
on PSD over frequency can be observed, which is characteristic of multipath effects.

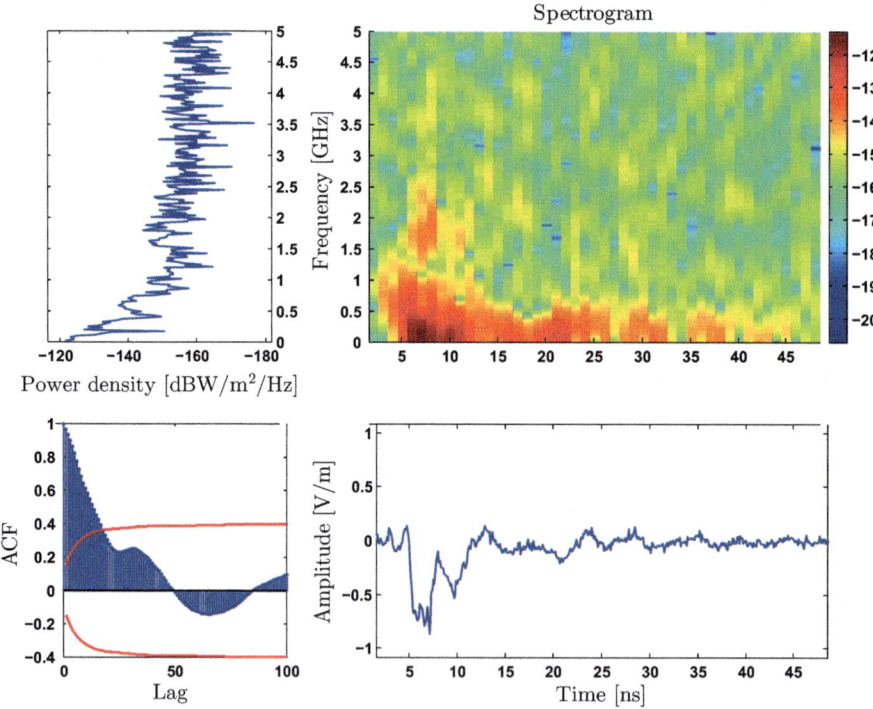

Fig. 5.1 Waveform measured from a 735 kV power transformer tank

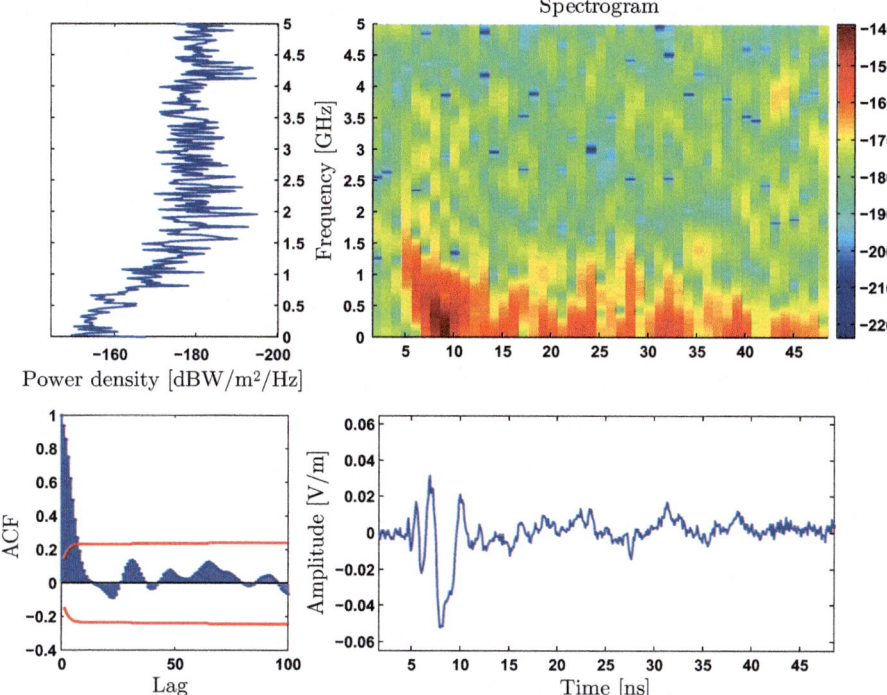

Fig. 5.2 Waveform measured from a current transformer tank

Moreover, due to their transient behaviour, there are significant correlations for high values of lag. As a result, using Middleton Class A and B and α-stable models might not be reasonable because an impulse is modeled by a single sample. These waveforms can be seen as colored impulsive noise. The autocorrelation function shows that there are significant correlations for large values of lag. The ACFs decay slowly and go under the confidence interval.

5.3.2   A Physical Interpretation

When a discharge occurs, the emission of electromagnetic radiations propagate through the environment. Let $aD(\bar{\theta}, t)$ be the original transient impulse induced by an elementary electric arc discharge where a is a random amplitude, $\bar{\theta}$ is a set of time-invariant random variables denoting the duration, time decay, phase shift, and other spectral characteristics. In the experimentations, the measured waveforms show that the amplitudes are distorted and attenuated. As a result, the received impulsive noise might be subject to fading. By denoting $W(\bar{\theta}, t)$ as the convolution product of the original transient impulsive waveform $D(\bar{\theta}, t)$ and the impulse response of the measurement setup, the resulting impulsive waveform distorted and attenuated by multiple reflected waves can be written as:

$$u(\theta, t) = aW(\bar{\theta}, t) * h(t) \tag{5.1}$$

where θ is a set of random parameters characterizing the duration, time decay, phase shift, and other spectral characteristics that have been distorted by the impulse response of the multipath channel $h(t)$. This can be written as:

$$h(t) = \sum_k \alpha_k e^{j\varphi_k} \delta(t - \tau_k) \tag{5.2}$$

where $\{\alpha_k\}$ is a real positive gain of the path k, $\{\varphi_k\}$ the associated phase shift and $\{\tau_k\}$ is the propagation delay. Equation (5.1) can be rewritten as a shot-noise process:

$$u(\theta, t) = \sum_k \alpha_k e^{j\varphi_k} aW(\bar{\theta}, t - \tau_k) \tag{5.3}$$

The multiple reflected waves induced by the environment cause both destructive and constructive interferences that are received by the antenna. Therefore, the impulsive waveforms are distorted by multipath effects, as seen in Figs. 5.1 and 5.2.

5.4   The Proposed Model

In this section, we propose a general model for impulsive noise induced by electric arc discharges when the waveforms are transient and their amplitudes are subject to multipath effect. In this chapter, we will provide the appropriate model for impulsive noise waveforms induced by an electric arc discharge. The goodness-of-fit of the proposed model will be analyzed to verify the its accuracy.

5.4.1   Theory of Filters and Its Relationship with Time Series Models

In this chapter, the modeled impulsive waveform is a time-series which is a series of data points indexed by a discrete time set T, a subset of all integers $\mathbb{Z}$. We choose to denote the stochastic process modeling an impulsive noise waveform $\{u_t\}_{t \in T \subset \mathbb{Z}}$. This notation is commonly used in literature [149–151]. In keeping with the model described by [141], the random process u_t can be modeled as a discrete time linear filter in the z-domain, such that:

$$U(z) = H_d(z)\varepsilon(z) \tag{5.4}$$

where $U(z)$ is the output of the filter, $H_d(z)$ is the z-transform of an impulse response of the original impulse waveform received at the antenna, and $\varepsilon(z)$ is the input stress characterizing the electromagnetic disturbance in the environment. $H_d(z)$ can be written as:

$$H_d(z) = \frac{1 + \sum\limits_{k=1}^{q} \psi_k z^{-k}}{1 - \sum\limits_{i=1}^{p} \phi_i z^{-i}} \tag{5.5}$$

where all coefficients ϕ_i and ψ_k for any $i = \{1, \cdots, p\}$ and $k = \{1, \cdots, q\}$ lie outside the unit circle. By replacing $H_d(z)$ in Eq. (5.5) into Eq. (5.4), we have:

$$U(z)\left(1 - \sum_{i=1}^{p} \phi_i z^{-i}\right) = \left(1 + \sum_{k=1}^{q} \psi_k z^{-k}\right)\varepsilon(z) \tag{5.6}$$

By using the inverse z-transform, we have in the time domain a time series by writing Eq. (5.6) as:

$$u_t = \sum_{i=1}^{p} \phi_i u_{t-i} + \varepsilon_t + \sum_{k=1}^{q} \psi_k \varepsilon_{t-k} \tag{5.7}$$

These terms within the equation induce correlated samples, since u_t depends on past samples. Furthermore, if the disturbance ε_t is a Dirac impulse, a deterministic transient impulse without any distortions is provided by the filter, as modeled in [141, 142]. This is not observed in the measurements. For ε_t to be modeled as a white Gaussian noise, its variance is constant over time, and the resulting waveform appears as a colored noise process spread indefinitely over time. This has not been observed here, and it is not appropriate for impulsive noise modeling either. Measured impulsive noise can be seen as colored noise; however, its variance might be not constant over time.

5.4.2 Definition of the Time Series Model

We define a lag operator L such that $L^i u_t = u_{t-i}$, where u_{t-i} is the real-valued sample of the process at the discrete time $t - i$. We also define the differential operator Δ such that $\Delta^i u_t = (1 - L)^i u_t$ is the i-th degree differencing operator. From the measured waveforms, it is convenient to assume that impulsive noises do not express any seasonal effect. From the Box-Jenkins methodology [145], non-stationary random processes can be trend-stationary, as written in Eq. (5.8) or difference-stationary, as written in Eq. (5.9), such that:

$$u_t = m_t + \varepsilon_t \tag{5.8}$$

$$\Delta^d u_t = m + \Psi(L)\varepsilon_t \tag{5.9}$$

where m_t is a deterministic mean trend and m is a constant mean, $\Psi(L) = 1 + \psi_1 L + \psi_2 L^2 \cdots$ is an infinite degree polynomial operator in which coefficients are absolutely summable, and all roots lying outside the unit circle, Δ^d are d-th degree differencing operators. By assuming a finite degree polynomial operator, we can define u_t as a generalized autoregressive integrated moving average ARIMA(p,d,q), modeled as:

$$\Phi_p(L)\Delta^d u_t = m + \Psi_q(L)\varepsilon_t \qquad (5.10)$$

where $\Phi_p(L) = 1 - \phi_1 L - \phi_2 L^2 \cdots - \phi_p L^p$ and $\Psi_q(L) = 1 + \psi_1 L + \psi_2 L^2 \cdots + \psi_q L^q$ are respectively the p-th and the q-th degree operator polynomials where all coefficients are absolutely summable and all roots lie outside the unit circle. The next task is to define the degrees (p,q) of the ARMA model and the degree d of differencing operator that fit the measurements. However, in ARIMA models, the process ε_t is assumed to be uncorrelated with a constant mean or a constant variance over time. In Sect. 5.3, measured impulsive noise may exhibit heteroskedasticity, i.e. non-constant variance. When the variance of ε_t is not constant over time, we can include additional conditional heteroskedasticity models to the time series u_t [152–154]. Among these models, we can mention autoregressive conditional heteroskedascity (ARCH), generalized ARCH (GARCH) or exponential GARCH (EGARCH). In these processes, the standard deviation of ε_t noted as σ_t or the variance noted as σ_t^2 is time-dependent. In this condition, we formally express ε_t by:

$$\varepsilon_t = \sigma_t \epsilon_t \qquad (5.11)$$

where ϵ_t is the white noise process. The time-dependent variance σ_t^2 can be modeled as an ARMA process for time-dependent effect of the disturbance ε_t [152, 153, 155]. We will define a suitable model, which fits measurements based on our assumptions. The latter can be tested with various tests to be defined. In this chapter, the analysis of the time series is restricted to the measured waveforms presented in Figs. 5.1 and 5.2. Similar conclusions with many measured impulsive noises can be drawn.

5.4.3   Tests for Unit Roots

A unit root is a process, in which the first difference of a time-series is stationary [144, 156]. Before estimating and selecting a model, it is necessary to test the "*stationarity*" (i.e. $d > 0$ [144]) and unit roots of the measured impulsive noises to avoid the problem of spurious regression [145, 149, 151]. To do so, we can treat the ARIMA(p,d,q) model as an ARMA($p + d,q$) model. As a result, we can write Eq. (5.10) as follows:

$$\Phi_{p+d}(L)u_t = m + \Psi_q(L)\varepsilon_t \qquad (5.12)$$

where $\Phi_{p+d}(L) = \Phi_p(L)(1 - L)^d$. Since measured impulsive noise fluctuates around zero-mean, we assume that $m = 0$. Next, since all roots of the polynomial $\Psi_q(L)$ lie outside of the unit circle, the invertibility condition of the model is satisfied and thus we have:

$$\Psi_q(L)^{-1}\Phi_{p+d}(L)u_t = \varepsilon_t \tag{5.13}$$

We assume that a polynomial $\Omega(L)$ exists related to polynomial $\Psi_q(L)^{-1}\Phi_{p+d}(L)$ such that Eq. (5.13) can be written as an AR process [156], defined by:

$$\Delta u_t = (\Omega(L) - 1)u_{t-1} + \varepsilon_t$$
$$\Delta u_t = \rho u_{t-1} + \sum_{i=1}^{k-1} \gamma_i \Delta u_{t-i} + \varepsilon_t \tag{5.14}$$

where γ_i is a function of the ARMA process and ρ is a coefficient. If the time series u_t is suspected to be non-stationary in difference, then a unit root exists. In this case, $\rho = 0$. In such a condition, the ARIMA($p,1,q$) model is recommended. The presence of unit roots may explain the non-stationary behaviour of the measurements. As a result, unit root tests can be used to determine whether impulsive noise contains a unit root.

The augmented Dickey-Fuller (ADF) [143] and the Phillips-Perron (PP) [144] tests can be used to assess the presence of a unit root in the impulsive waveform. Based on the model defined in Eq. (5.14), without any intercept or a drift terms, the null hypothesis of the unit root is $H_0 : \rho = 0$ against the alternative hypothesis, $H_1 : \rho < 0$. The estimation of $\hat{\rho}$ is based on ordinary least squares (OLS) and the test statistics for the null hypothesis is given by the t-statistic. However, the test statistics do not follow a standard distribution. Hence, the limiting distributions have been derived [151]. The resulting value is compared to the interpolated Dickey-Fuller critical values, based on the tables in [150] for the decision rule. An approximation of the p-values is given in [157] on the basis of a regression surface. However, for the ADF test, the process ε_t should be stationary; in other words, ε_t does not exhibit conditional heteroskedasticity. Furthermore, the selected number of lags should be appropriate to keep the test unbiased or to decrease the power of the test.

At the same time, the PP test is a non-parametric test, based on the same model used in ADF tests, that modifies the ADF test statistics. It is made robust to serialize correlation and heteroskedasticity in ε_t [144, 158]. The PP test procedure remains the same as the ADF test, and it uses the critical values based on the same tables in [150]. The tests need the number of autocovariance lags to include in the Newey-West estimator of the long-run variance, which depends on the number of observations M [158]. This number is given by $\lfloor 4 (M/100)^{2/9} \rfloor$ where $\lfloor \cdot \rfloor$ is the floor function. Another advantage is that we do not need to specify the number of lags for the test. Hence, Eq. (5.14) for the PP test is reduced to:

$$\Delta u_t = \rho u_{t-1} + \varepsilon_t \tag{5.15}$$

Table 5.1 Results of PP test obtained from the measured impulsive noises

Waveforms	Nb. obs	Test-stat.	Crit. val	$\hat{\rho}$
PTT	510	-2.68	-1.94	-0.027
CTT	500	-4.27	-1.94	-0.062

The samples in the measurements may exhibit heteroskedasticity. As a result, it is reasonable to choose the PP test. Table 5.1 depicts the results of the PP test. The waveforms, which are presented in Figs. 5.1 and 5.2 are tested. The test is set for a significance level of 5%. The test statistic values in the table are all below the interpolated Dickey-Fuller critical values and the estimated $\hat{\rho}$ is negative and the p-values are closed to zero. As a result, the test indicates the rejection of the unit root null in favor of the alternative model, at a 95% confidence interval. In other words, measured impulsive waveforms do not need to be differentiated. Thus, we should use ARMA(p,q) models with no integration for transient impulsive noise modeling.

5.4.4 Estimation and Selection

Since data do not need to be differentiated, we can use the ARMA(p,q) models. The next task is to determine the degree (p,q) of the model at which coefficients ϕ_i and ψ_k can be estimated from the observed data u by using the maximum likelihood estimation (MLE). The model can be chosen based on the Akaike Information Criterion (AIC) [146, 159] or the Schwarz Bayesian Information Criterion (SBIC) [147], written respectively as:

$$\text{AIC} = 2\kappa - 2\log(\mathfrak{L}(\boldsymbol{\Theta}_\kappa|u)) \tag{5.16a}$$

$$\text{SBIC} = -2\log(\mathfrak{L}(\boldsymbol{\Theta}_\kappa|u)) + \kappa\log(M) \tag{5.16b}$$

where $\mathfrak{L}(\boldsymbol{\Theta}_\kappa|u)$ is the optimized likelihood objective function value of the proposed model parametrized by a vector $\boldsymbol{\Theta}_\kappa$, to be estimated from the observed data u. The number of free parameters to be estimated is denoted by κ, and M, the number of observations. These values include a penalty for complex models with additional parameters. The SBIC is more severe than the AIC in terms of penalties, due to its relationship with the number of observations M.

Note that the model with the lowest AIC or SBIC has the best fit. It has been found that the ARMA(7,2) for PTT and the ARMA(4,1) for CTT are suitable models because AIC and SBIC values are small compared to the other values of p and q. The residuals of fitted ARMA models can be inferred to check the goodness-of-fit.

5.5 The Goodness-of-Fit

In this section, we study the models' goodness-of-fit. The residuals of fitted ARMA models are analyzed to check the adequacy of the model. We will show that the residuals express conditional heteroskedasticities. Therefore, ARMA models need to be improved by adding ARCH effects.

5.5.1 Analysis of the Residuals

Residuals are helpful for modeling the disturbance term ε_t so that u_t behaves like transient impulsive noise with correlated samples. It is convenient to analyse the standardized residuals of fitted ARMA models by checking their whiteness. This can be showed by using the normal probability plot in which probability distributions of residuals can be graphically compared to the normal distribution [127]. In addition, a Kolmogorov-Smirnov (KS) test can be used. These allow us to assess whether residuals could come from a normal distribution. Moreover, ACF and partial ACF (PACF) can be plotted to check whether residuals are uncorrelated. The bounds for autocorrelation set at 95% of confidence are given by:

$$B = \pm \frac{1.96}{\sqrt{M}} \qquad (5.17)$$

The analysis of the autocorrelation can be refined by using the portmanteau test given by Ljung-Box Q-test for Residual Autocorrelation [160, 161]. It tests for autocorrelation at multiple lags jointly. The null hypothesis is that the first K autocorrelations are jointly zero. For a number of observations M, the Ljung-Box Q-statistic is given by the portmanteau statistic written as:

$$Q_\varepsilon = M(M+2) \sum_{k=1}^{K} \frac{\hat{r}_k^2(\varepsilon)}{M-k} \qquad (5.18)$$

where $\hat{r}_k^2(\varepsilon)$ is the estimated autocorrelation of residuals at lag k. Under the null hypothesis, Q_ε follows a chi-square distribution χ^2_{K-p-q} asymptotically with $K - p - q$ degrees of freedom, which depends on the (p,q) of the selected ARMA models. If a p-value is greater than a given significance level, then the test fails to reject the null hypothesis and proves that the residuals are not autocorrelated.

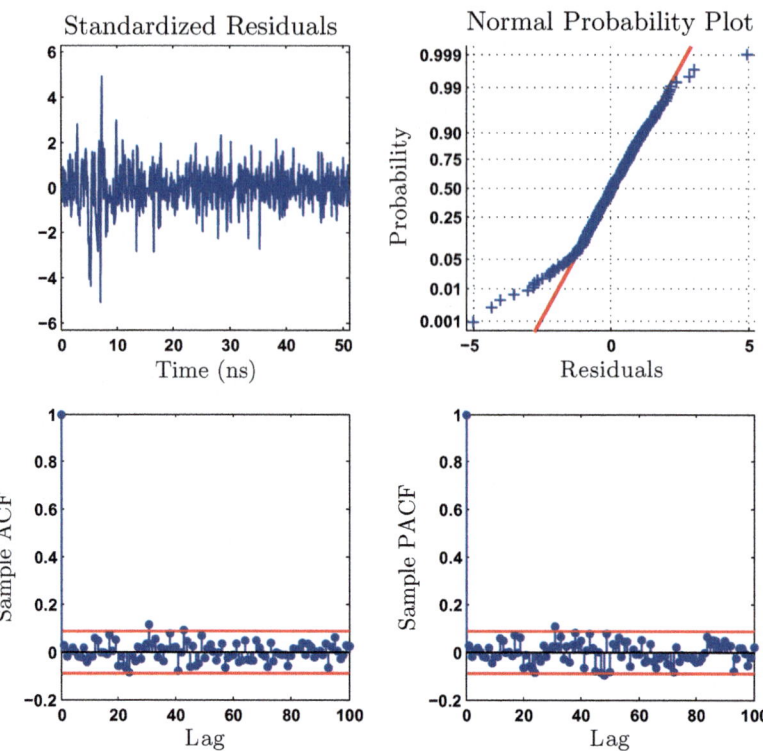

Fig. 5.3 Analysis of the residuals of fitted ARMA(7,2) model

5.5.1.1   Residuals of Fitted ARMA(7,2)

The distribution of the residuals is heavy-tailed since a deviation at the extreme values is observed in the normal probability plot graph in Fig. 5.3. This graph allows for the identification of deviations such as skewness or excess kurtosis compared to the normal distribution [127, 162]. The p-value of the KS test is below 0.05. Thus, the test rejects the null hypothesis of normal distribution at a 95% confidence interval. Therefore, the distribution could not come from a normal distribution since the curve is not linear, and the KS test rejects the null hypothesis. Furthermore, on the ACF and PACF, the correlation coefficients for different values of lags are, in general, below the limits for autocorrelation. Therefore, the residuals might be uncorrelated. In order to check whether residuals are uncorrelated, we can use the Ljung-Box Q-test for residual autocorrelation. The test is set to a significance level of 5% with different values of K-first autocorrelations. Results of the test are depicted in Fig. 5.4, in which the critical values are plotted in a red curve and the calculated Q-statistic values are plotted in a blue curve for the different values of K-first autocorrelations. The p-values are represented in this figure. These values are compared to 0.05, since the test is set to a significance level of 5%.

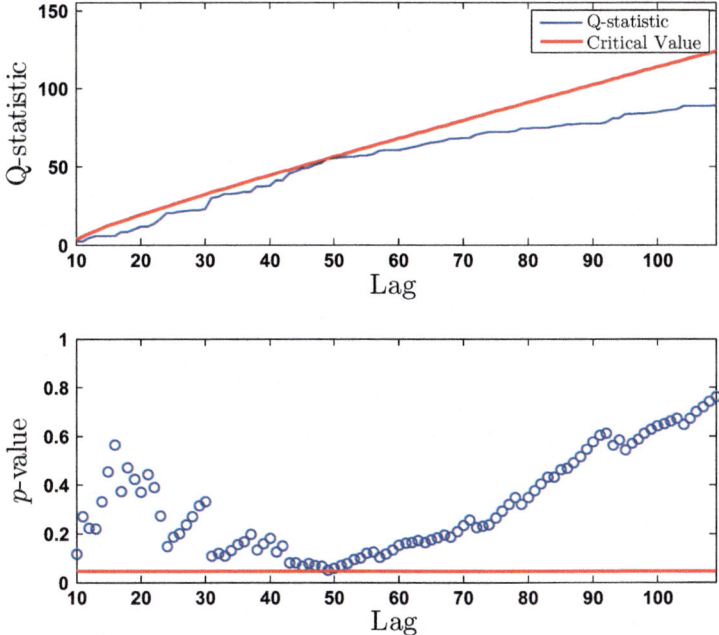

Fig. 5.4 Results of Ljung-Box Q-test for residual autocorrelation of the ARMA(7,2)

We can see that all values of the Q-statistic are below the critical values for different values of K. The p-values are all above 0.05. As a result, the test fails to reject the null hypothesis that the residuals are not autocorrelated at a confidence interval of 95% for different values of K-first autocorrelations tests.

5.5.1.2 Residuals of Fitted ARMA(4,1)

The distribution of the residuals is less heavy-tailed as observed in the normal probability plot graph. The KS test fails to reject the null that the distribution comes from the normal distribution. Nevertheless, a deviation at the negative values is observed, as in Fig. 5.5. The distribution might be left-skewed. Moreover, some correlation coefficients are above the autocorrelation bounds in the ACF and the PACF. Hence, we may not conclude graphically whether the residuals are uncorrelated.

The Ljung-Box Q-test for residual autocorrelation is also set for a significance level of 5% with different values of K-first autocorrelations. The results of the test are depicted in Fig. 5.6. We can see that there is no evidence to reject the null hypothesis that proves that the residuals are not autocorrelated at a 95% confidence interval, since all values of the Q-statistic are below the critical values and the p-values are all above 0.05 for different values of K-first autocorrelations.

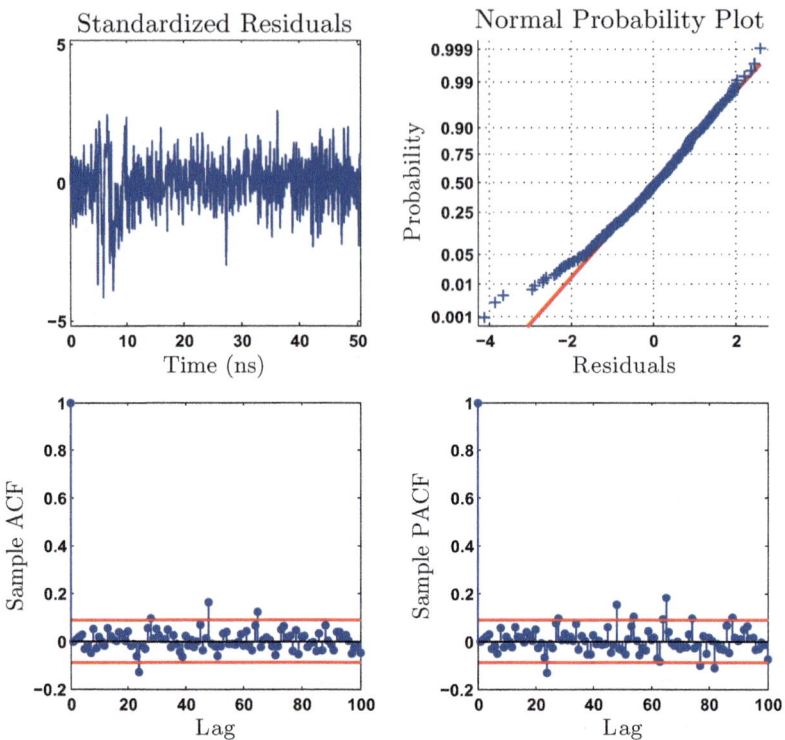

Fig. 5.5 Analysis of the residuals of fitted ARMA(4,1) model

In summary, for both models ARMA(7,2) and ARMA(4,1), the distributions of the residuals do not come from a normal distribution and the Ljung-Box Q-tests fail to reject the null hypothesis that the residuals are not autocorrelated. We can conclude that the disturbance terms ε_t are uncorrelated non-Gaussian noise.

These proposed models need to be improved by adding additional parameters. Increasing the number of (p,q) can refine the models. However, the models would be more complex and overfitting problems can occur [163]. Moreover, the heavy-tailed distributions of the residuals cannot be explained by increasing the number of (p,q). On the other hand, the proposed models can be refined by assuming that these residuals are the consequence of non-linear effects such as heteroskedasticities [152, 155].

5.5.2 Tests for Heteroskedasticity

The ARMA(p,q) process with unconditional variance of the disturbance process ε_t may not be a suitable assumption for modeling impulsive noises. When the process

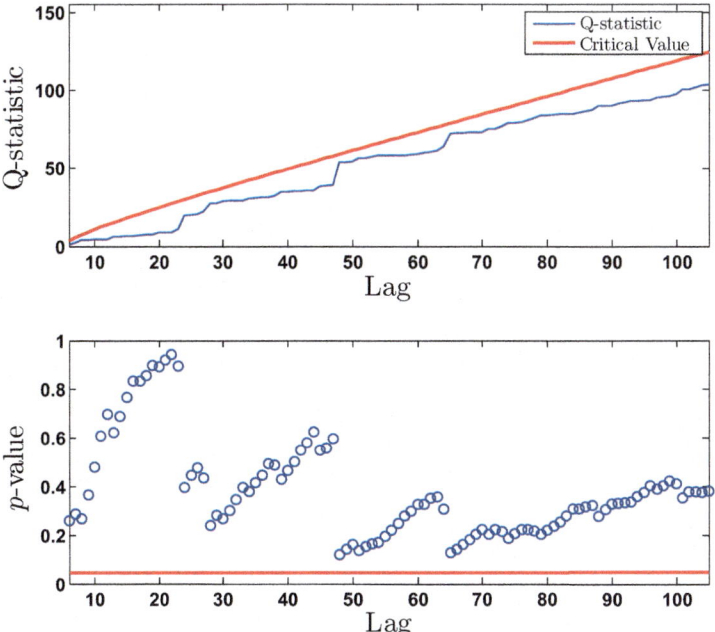

Fig. 5.6 Results of Ljung-Box Q-test for residual autocorrelation of the ARMA(4,1)

contains ARCH effects, we can write the variance σ_t^2 of the process ε_t given the information set available, which is denoted by $\mathscr{I}_{t-1} = \{\varepsilon_{t-1}, \varepsilon_{t-2}, \cdots\}$ at discrete time $t-1$ [152, 153, 155]:

$$
\begin{aligned}
\sigma_t^2 &= \mathbb{E}[\varepsilon_t^2 | \mathscr{I}_{t-1}] \\
&= \Upsilon_{q'}(L)\varepsilon_t^2 \\
&= \upsilon_0 + \sum_{i=1}^{q'} \upsilon_i \varepsilon_{t-i}^2
\end{aligned}
\tag{5.19}
$$

where $\mathbb{E}[\cdot]$ denotes the expectation and υ_i are coefficients of the polynomial $\Upsilon_{q'}(L)$. We assume that ε_t is zero-mean. The assumption for conditional heteroskedasticity can be tested by the Ljung-Box Q-test of the squared residuals of fitted ARMA models [164]. In addition, the Engle's ARCH test [153] for conditional heteroskedasticity can be set to ensure that the measurements contains ARCH effects.

Several authors [164, 165] have noted that the analyses of the squared residuals are useful for the detection of non-linear types of statistical dependence in the residuals of fitted ARMA models. Similarly to the Ljung-Box Q-statistic in

Eq. (5.16), the test statistic of squared residuals denoted by ε^2 is given by the following portmanteau statistic:

$$Q_{\varepsilon^2} = M(M+2) \sum_{k=1}^{K} \frac{\hat{r}_k^2(\varepsilon^2)}{M-k} \tag{5.20}$$

Under the null hypothesis, Q_{ε^2} asymptotically follows a chi-square distribution χ_K^2 with K degrees of freedom [164]. The test rejects the null hypothesis of no autocorrelation in the squared residuals if $Q_{\varepsilon^2} > \chi_K^2$ where χ_K^2 is the chi-square distribution table value.

In addition, the Engle's ARCH test can be used. Following Eq. (5.19), the null hypothesis is $H_0 : v_i = 0$ for all $i = 1 \cdots q'$. The test used is the Lagrange multiplier (LM) statistic MR^2, where M is the sample size and R^2 is the coefficient of determination from fitting the ARCH(q') model for a number of lags q'. The null hypothesis indicates the failure to reject the no ARCH effects [153]. Under the null hypothesis, the asymptotic distribution of the test statistic is χ^2 with q' degrees of freedom. Note that the test can be applied to GARCH models since a GARCH(p',q') model is locally equivalent to an ARCH($p' + q'$) model [152].

As depicted in Figs. 5.7 and 5.8, the Ljung-Box Q-test for squared residual autocorrelation rejects the null hypothesis that the squared residuals are not autocorrelated at 95% confidence for different values of K-first autocorrelations. As a result, a non-linear effect should be taken into account in the ARMA(p,q) models.

Moreover, as seen in Figs. 5.9 and 5.10, the null hypothesis is rejected for different values of lags q' at a confidence interval of 95%. Note that the test fails to reject the null hypothesis of no ARCH effects in the residuals of fitted ARMA(4,1) models at lags up to $q' = 75$. Due to its complexity, in practice we will not use an ARCH model with q' up to 75. As a result, we can argue that the measurements contain heteroskedasticities in conformity with results obtained from the Ljung-Box Q-tests for square residual autocorrelation.

5.5.3   Analysis of the Residuals of the Improved Models

Since the residuals of fitted ARMA(p,q) models contain conditional heteroskedasticities, the models can be refined by adding a non-constant variance in the disturbance term ε_t. It is believed that the variance is a form of the power law decay function. Hence, the EGARCH models proposed by [154] might be appropriate. According to the AIC and SBIC, we find that ARMA(7,2)-EGARCH(9,6), denoted by *Model* 1 and ARMA(4,1)-EGARCH(12,8), denoted by *Model* 2, are suitable models for the measurements. We can now analyse the standardized residuals of fitted models.

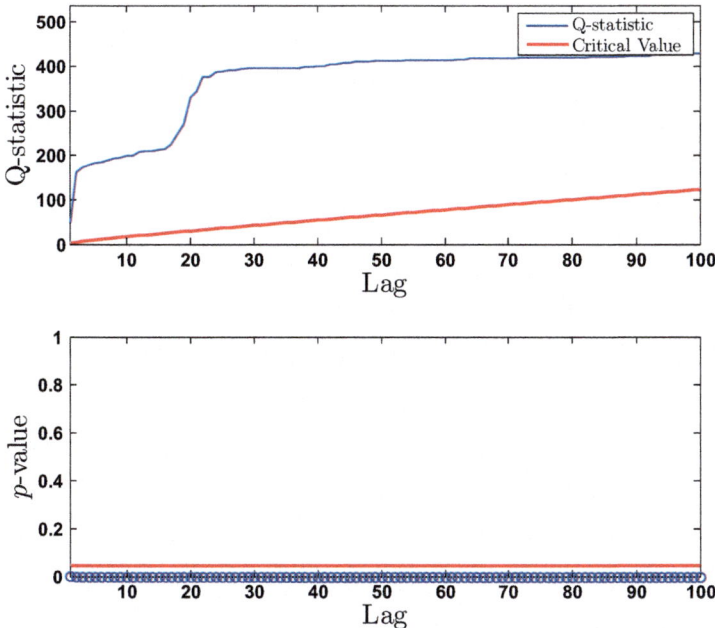

Fig. 5.7 Results of Ljung-Box Q-test for squared residual autocorrelation of the ARMA(7,2)

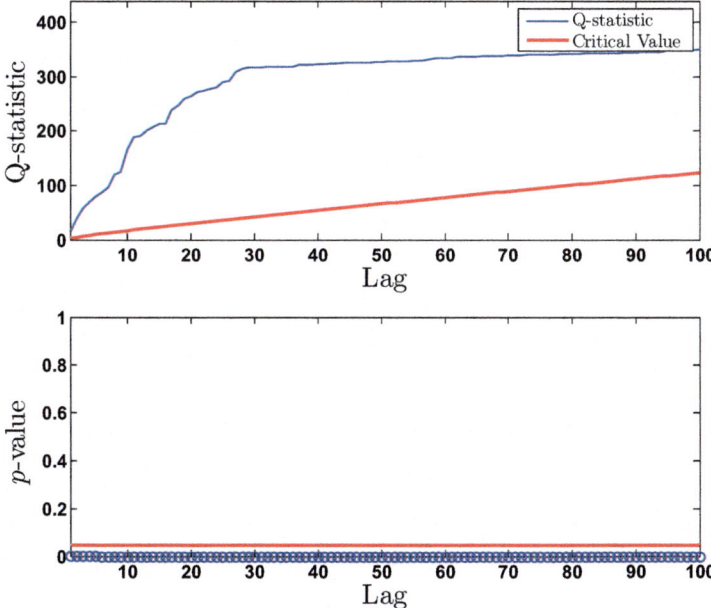

Fig. 5.8 Results of Ljung-Box Q-test for squared residual autocorrelation of the ARMA(4,1)

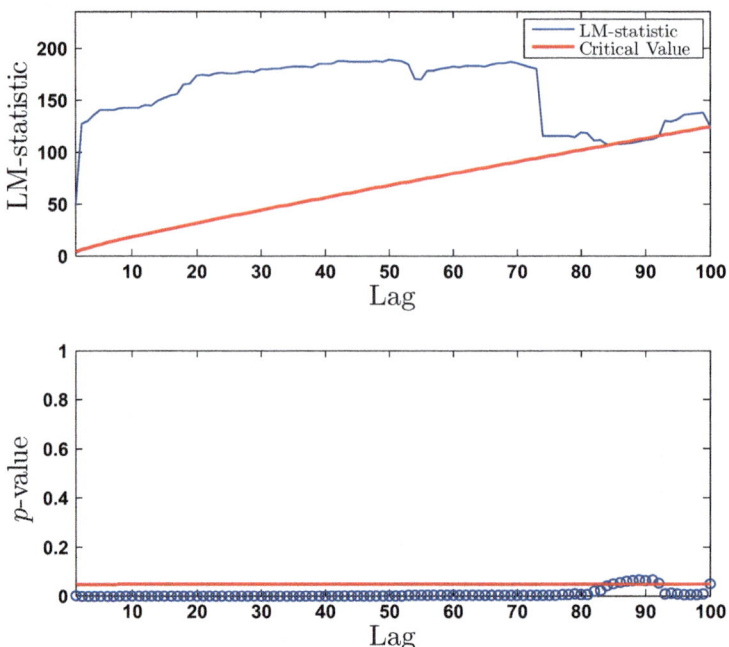

Fig. 5.9 Results of Engle test for residual heteroskadasticity of the ARMA(7,2)

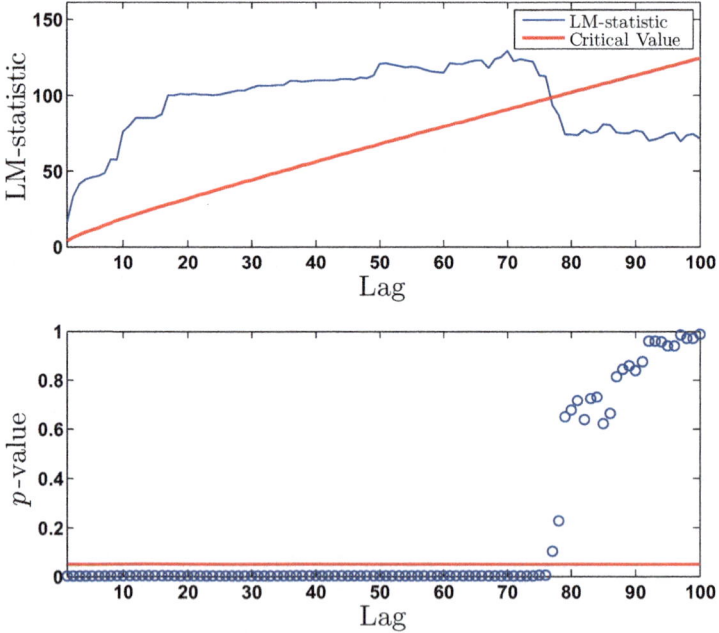

Fig. 5.10 Results of Engle test for residual heteroskadasticity of the ARMA(4,1)

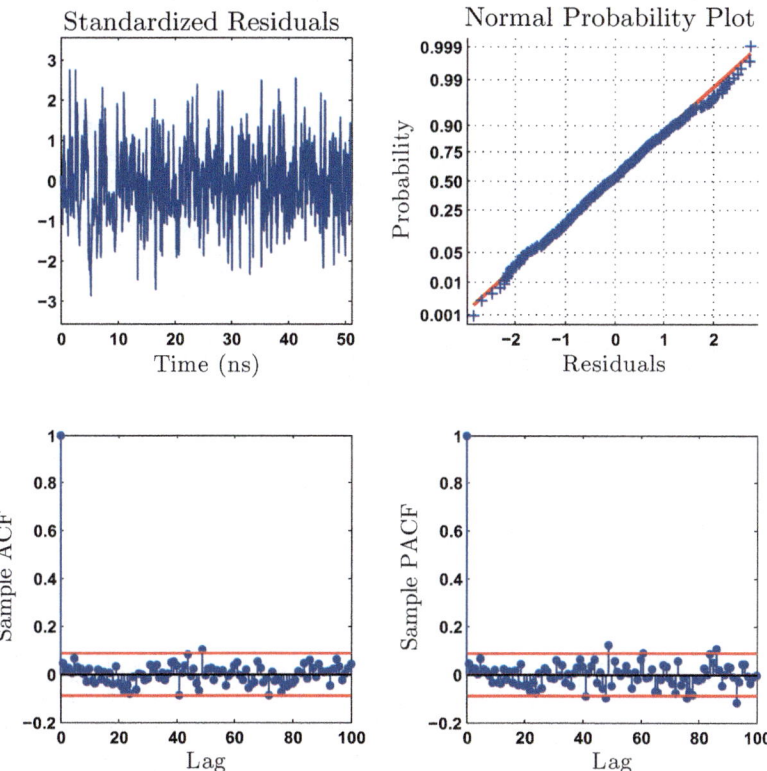

Fig. 5.11 Analysis of the residuals of fitted *Model* 1

In Figs. 5.11 and 5.12, the distribution of the residuals in each model are less heavy-tailed, not skewed and fit the red curve. In addition, the KS test's *p*-values are 0.9 and 0.48 for *Model* 1 and *Model* 2 respectively. Under such conditions, the residuals could come from a normal distribution. The ACF and the PACF show that many lags are far below the limits for autocorrelation. As a result, they might be uncorrelated, and we may conclude that the residuals might be white Gaussian noise.

By using the Ljung-Box Q-test for residual and squared residual autocorrelations, the results of the tests are resumed for a few values of K-first correlations in this chapter. As depicted in Tables 5.2 and 5.3, all *p*-values are above the significance level of 5%. Hence, the Ljung-Box Q-test fails to reject the null hypothesis that proves that the residuals are not autocorrelated at 95% of confidence for different values of K-first autocorrelations. Since we want a generic model, it is reasonable to approximate the residuals from fitted ARMA-EGARCH models as white Gaussian noise.

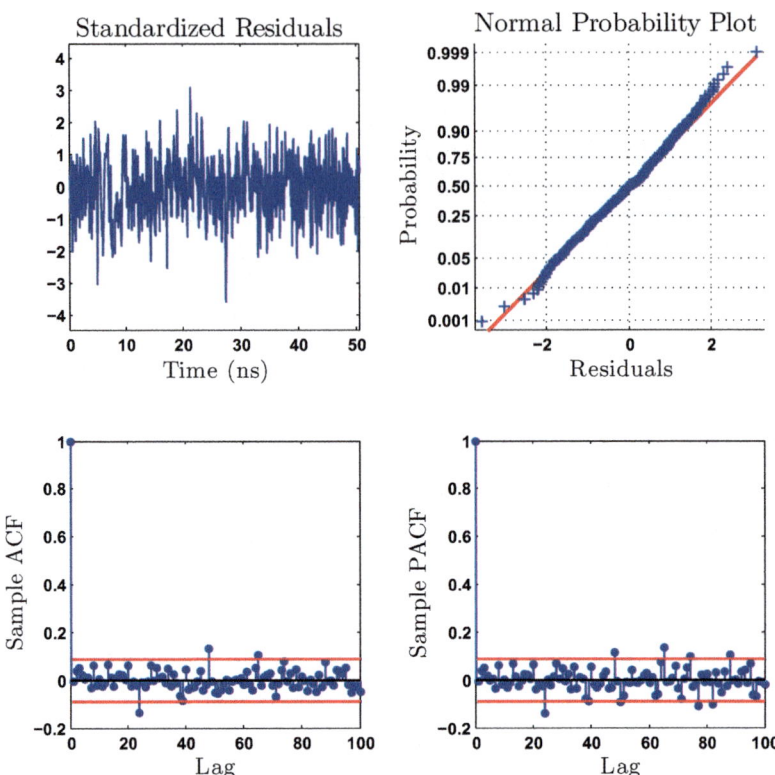

Fig. 5.12 Analysis of the residuals of fitted *Model* 2

Table 5.2 Results of Ljung-Box Q-test for residual autocorrelation

p-Value of Q statistic	$K = 10$	$K = 25$	$K = 60$	$K = 80$
Model 1	0.83	0.93	0.87	0.82
Model 2	0.96	0.96	0.98	0.98

Table 5.3 Results of Ljung-Box Q-test for squared residual autocorrelation

p-Value of Q statistic	$K = 10$	$K = 25$	$K = 60$	$K = 80$
Model 1	0.81	0.52	0.57	0.68
Model 2	0.47	0.24	0.28	0.2

5.5.4   Summary

Impulsive noise with correlated samples can be analyzed and modeled using an LTI filters approach. Their relationship with discrete time series can help us to choose a suitable filter, such as the number of parameters, and the input time series is given by ε_t. From this analysis, we can retain that:

- impulsive noise that is obtained by measurements can be fitted by ARMA(p,q) models where all coefficients may lie outside the unit circle. Thus, the stationarity and the invertibility conditions are satisfied;
- the impulsiveness of the data comes from conditional errors. Indeed, the residuals from fitted ARMA models contain ARCH effects as shown with the Ljung-Box Q-test for squared residual autocorrelation and the Engle test for residual heteroskedasticity;
- the ARMA models can be improved by adding ARCH effects such as EGARCH models. Thus, the standardized disturbance term $\epsilon_t = \varepsilon_t / \sigma_t$ can be approximated by a white Gaussian noise. This suitable approximation can reproduce distortions induced by the multipath propagation effects.

5.6 Simulation and Results

5.6.1 Simulation Parameters

In order to demonstrate the efficiency of the proposed model, we can simulate the resulting waveforms obtained at the output of the LTI filters. They can be compared to measured impulsive waveforms to determine the goodness-of-fit. The analysis of the obtained results is provided.

By using the ARMA(p,q) models fitted from the measurements, we define a disturbance term $\varepsilon_t = \sigma_t \epsilon_t$, as defined in Eq. (5.11) where ϵ_t is a white noise with normal distribution $\sim \mathcal{N}(0, 1)$ and σ_t is the time-dependent standard deviation. We choose to define the time-dependent function given by:

$$\sigma_t = \frac{\sigma_0}{t \nu \sqrt{2\pi}} \exp\left(-\frac{(\log t - \mu)^2}{2\nu^2}\right) \tag{5.21}$$

where ν and μ are parameters related to the rise and decay time and σ_0 is a normalized scale parameter. These parameters are adjusted to compare measurement and simulation results.

5.6.2 A Comparison of Measurement vs. Simulation Results

By using the ARMA(p,q) models fitted from the measurements, and by considering the disturbance term ε_t as a white Gaussian noise whose variance is a time-dependent function, we can plot the results obtained by a simulation. Simulated waveforms and spectral densities can be compared to measured impulsive noise.

We use the ARMA(7,2) model with conditional heteroskedasticity. It is denoted by *Model* 1. In Fig. 5.13, the resulting waveform obtained by the model is an

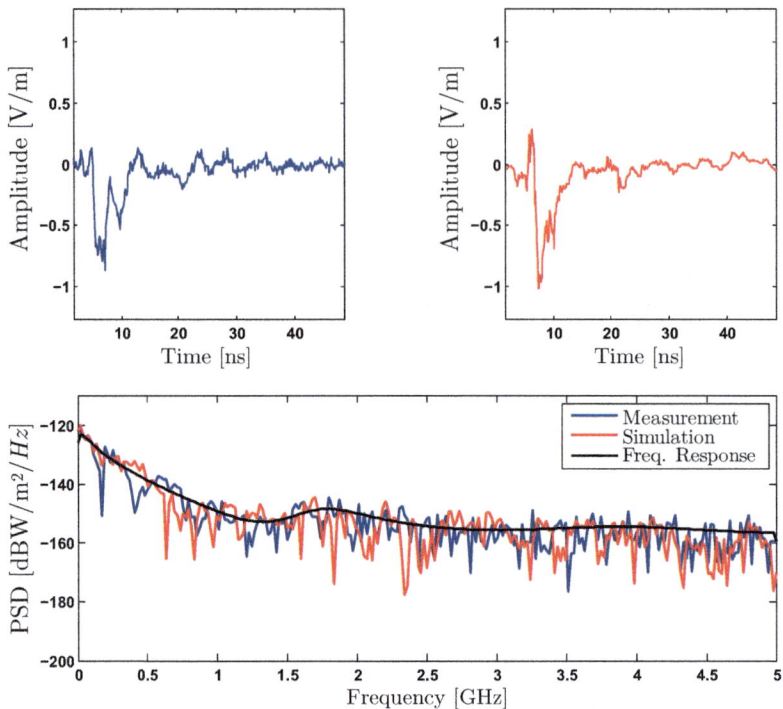

Fig. 5.13 Comparison of waveforms: PTT vs. *Model* 1

impulsive waveform. Its power spectral density, in red curve, accurately fits the PSD of measurements, which are in blue curve. The frequency response of the LTI filter is represented by the black curve. The PSDs are decaying as a form of $\sim f^{-\gamma}$ where $\gamma > 0$. The effect of the disturbance term induces local variations of the PSD.

For *Model* 2, we use the ARMA(4,1) model, with conditional heteroskedasticity. The obtained waveform and the PSD are depicted in Fig. 5.14. The waveform is impulsive and randomly distorted. By comparing the PSDs, the simulation fits the measurement accurately.

5.6.3   Analysis of Simulated Impulsive Waveforms

A more detailed analysis of simulated waveforms can be provided by studying their spectrograms and their autocorrelation functions. As depicted in Figs. 5.15 and 5.16, the obtained results are quite similar to the measured impulsive noises presented in Figs. 5.1 and 5.2. The samples of these impulsive waveforms are correlated. Indeed, slow decays, quite similar to the measurements, are observed in the autocorrelation functions. Hence, there are significant correlations at different

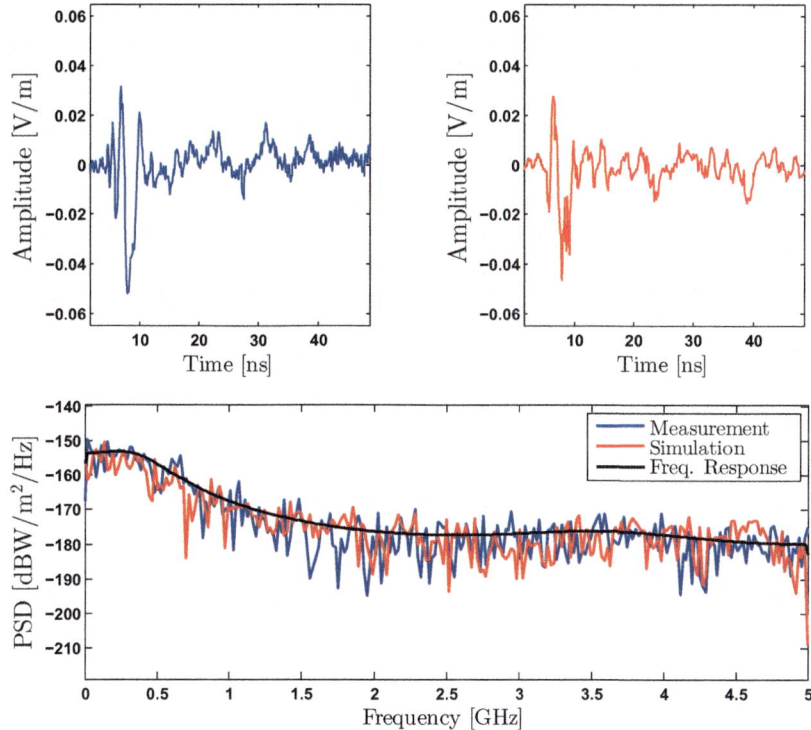

Fig. 5.14 Comparison of waveforms: CTT vs. *Model* 2

values of lags. Moreover, the effect of the disturbance term, modeled as a white noise with conditional variance, can reproduce the impulsiveness of the process. The spectrograms show that the power density of the simulated impulsive waveforms are strong for a short duration with a large frequency bandwidth, and it decreases over time and frequency. Note that the proposed model is based on the modeling of second-order statistics of the measurements. The model can be validated using second-order statistics such as PSD, spectrogram and ACF. This approach is accurate for modeling these impulsive noises.

5.6.4 Advantages and Limitations of the Proposed Model

Compared to partitioned Markov chain based-model, the proposed approach allows for an accurate estimation of the spectral characteristics of PD from data via a simple and straightforward estimation procedure [143–147]. In addition, the measure of the goodness-of-fit allows us to assess the adequacy and the accuracy of time series models.

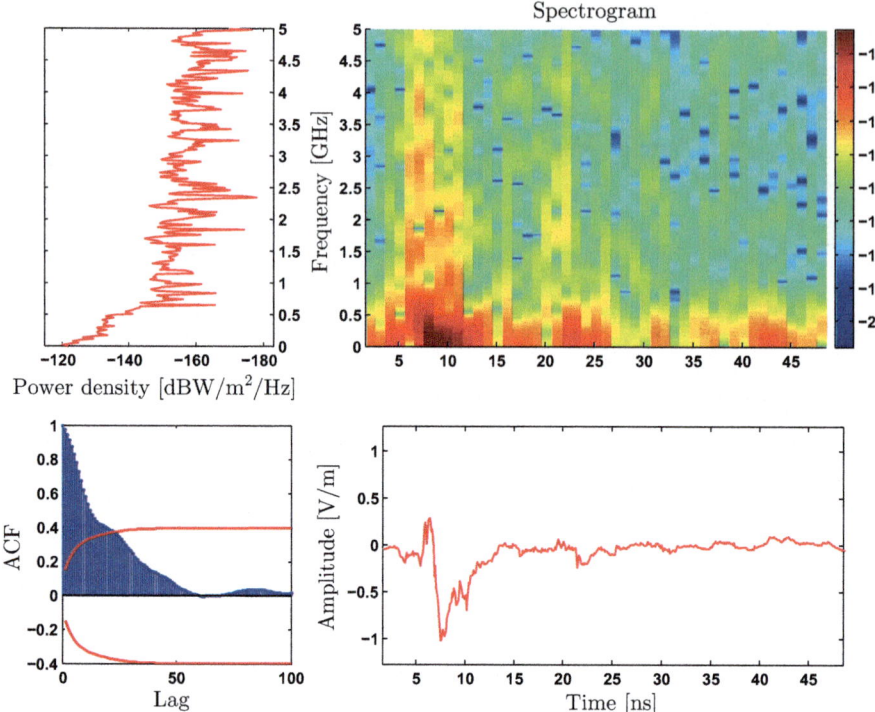

Fig. 5.15 Waveform obtained with the *Model* 1

The main issue of such approaches is the selection of the time series model and the number of parameters to be estimated. The complexity of the model increases as the number of parameters increases and overfitting problems can occur [163]. There is a clear trade-off between the accuracy and the complexity of such models. Fortunately, this can be resolved by using the Akaike information criterion [146, 159] or the Schwarz Bayesian information criterion [147].

5.7   Conclusion

In this chapter, we propose a new approach to capturing spectral characteristics of EMIs induced by PD activity, based on second-order statistics EMIs from PD activity. Experimentations and measurement campaigns show that the wideband impulsive noise waveforms emitted by electric arc discharges are transient when samples are correlated. Moreover, their power spectral densities have a form of approximately $\sim f^{-\gamma}$ where $\gamma > 0$ is an arbitrary exponent.

These impulsive noises can be modeled by an LTI filter in which correlated samples are induced by the autoregressive and/or the moving average terms. We can

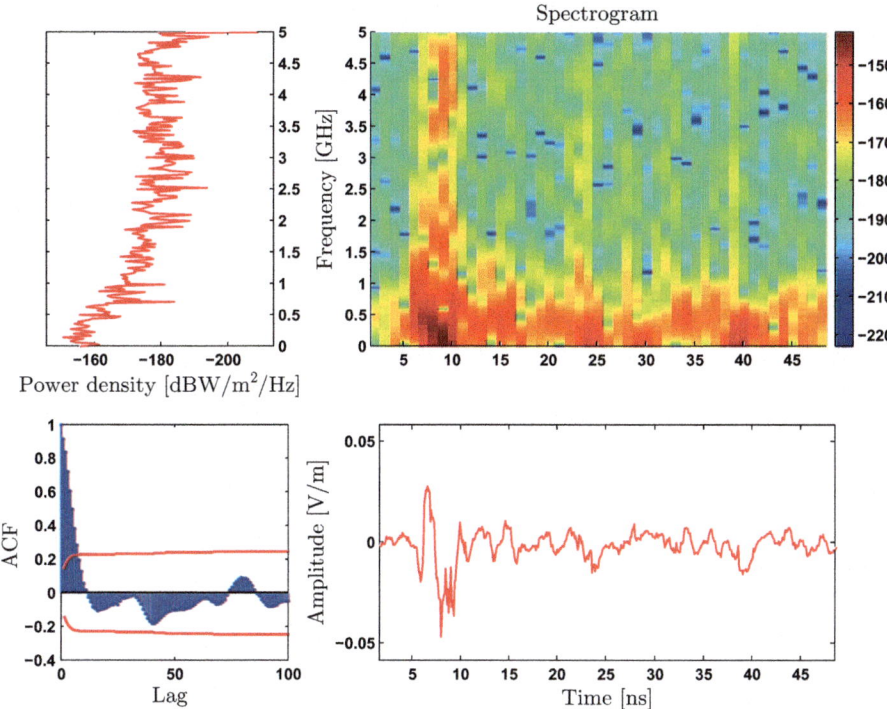

Fig. 5.16 Waveform obtained with the *Model* 2

analyse the filter in a time domain by using a time series in which all coefficients, the AR(p) and the MA(q) terms, can be estimated by using the maximum-likelihood function.

From the analysis of the time series fitted from the measurements, all coefficients of the ARMA(p,q) models lie outside the unit circle in which the stationarity condition is ensured. The residuals of fitted ARMA models contain ARCH effects in which the variance could be seen as a time dependent function, i.e. not constant over time. As a result, the models can be refined by adding heteroskedasticity in the disturbance term. The standardized residuals from fitted ARMA models with heteroskedasticity can be approximated by a white noise. Consequently, the disturbance terms can be modeled as white noise with a time dependent variance. It allows for the reproduction of distortions induced by the multipath propagation effects. The efficiency of the approach is demonstrated by comparing the waveforms obtained by simulation to those seen in measurements. Their second-order statistics fits the measurements accurately.

In future work, the proposed model can be extended to vector ARMA models with heteroskedasticity in which the vector is a collection of many measured impulsive noise waveforms. The estimation and the analysis follow the same procedure as described in this chapter exactly. Under this condition, the ARMA(p,q) models and the parameters defining the heteroskedasticity can be estimated empirically from any discharge sources generated by any HV equipment in substation environments such as power transformers, overhead power lines, or circuit breakers.

Chapter 6
Wideband Statistical Model for Substation Impulsive Noise

There exist other possibilities of introducing correlation as compared to impulse response models of Chap. 5. We propose in this chapter an impulsive noise model using a Partitioned Markov Chain (PMC) in order to generate samples with a correlation that produces impulses with a damped oscillating waveform.

6.1 Introduction to PMC Model

The concept of partitioned Markov chain has been introduced by Zimmermann [23] for representing the impulse occurrence in PLC communications. This concept is modified by associating to each state a Gaussian distribution carefully parameterized, which provides a more realistic distribution of the samples, including the time correlation. Up to 19 states is used: one to represent the background noise and up to 18 states to represent the impulse samples. The Markov chain is configured in order to generate impulses with correlated samples that can produce PDFs of impulsive noise characteristics that are more similar to measurements, especially in terms of the power spectrum of the impulses.

For a better understanding of the work, we explain first, based on substation measurements, how to interpret the events and the characteristics of the impulses; thereafter we present the configuration of our partitioned Markov chain. In Chap. 2, it is noticed that a correlation might exist between the duration and the amplitude of the impulses. The observation of these two variables has provided a Pearson correlation coefficient of 0.85, which confirmed our assumption: the larger the impulses are, the longer they last.

Another particularity observed during the measurement campaign is that the impulses are sometimes grouped every 8.3 ms (Fig. 6.1), which represents one half of a 60 Hz cycle. However, the power spectrum of the impulses is not affected by the impulses being grouped with a 60 Hz cycle (Fig. 6.2). According to [35], the

© Springer International Publishing AG, part of Springer Nature 2019   119
B. L. Agba et al., *Wireless Communications for Power Substations:*
RF Characterization and Modeling, Wireless Networks,
https://doi.org/10.1007/978-3-319-91328-5_6

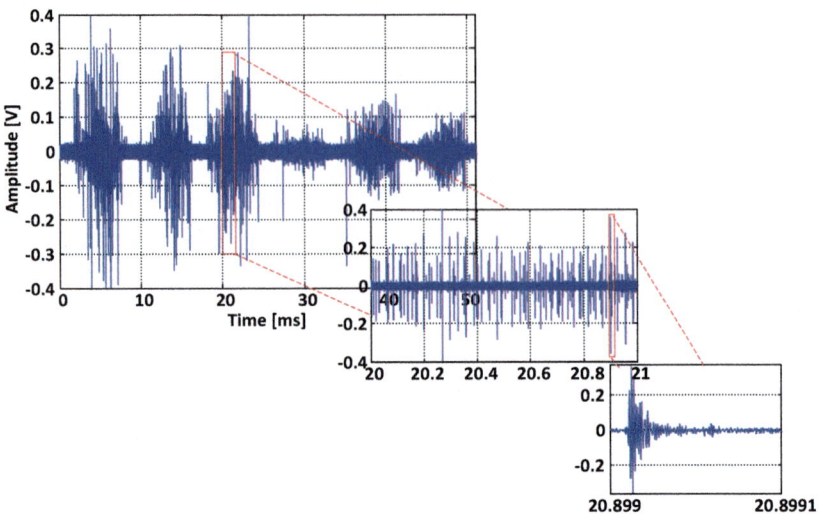

Fig. 6.1 Illustrative impulsive noise measurement in a 735 kV substation, band: 780 MHz to 2.5 GHz [118, 166]

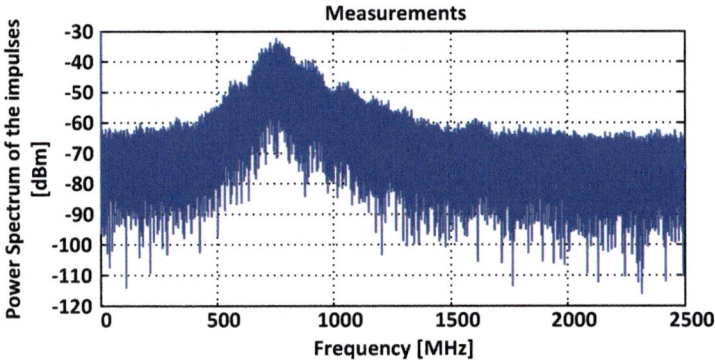

Fig. 6.2 Power spectrum of the impulses in a 735 kV substation, band: 780 MHz to 2.5 GHz [118, 166]

partial discharges that produce impulsive noise are created as soon as the electrical potential in the 60 Hz cycle reaches a critical value of the insulator; this critical value can be reached several times during a cycle and it is more likely that partial discharges occur around the maxima (positive and negative) of the 60 Hz cycle [35]. The correlation between the 60 Hz and the groups of impulse is however not always observed in our measurements, especially in substations where three-phase is used in the electric grid. Other measurements present an impulsive noise where there is no obvious correlation between the groups of impulse and the 60 Hz (Fig. 6.3).

Based on observations, we consider three main points. First, the background noise is constantly present in substations, however it is no longer observable when

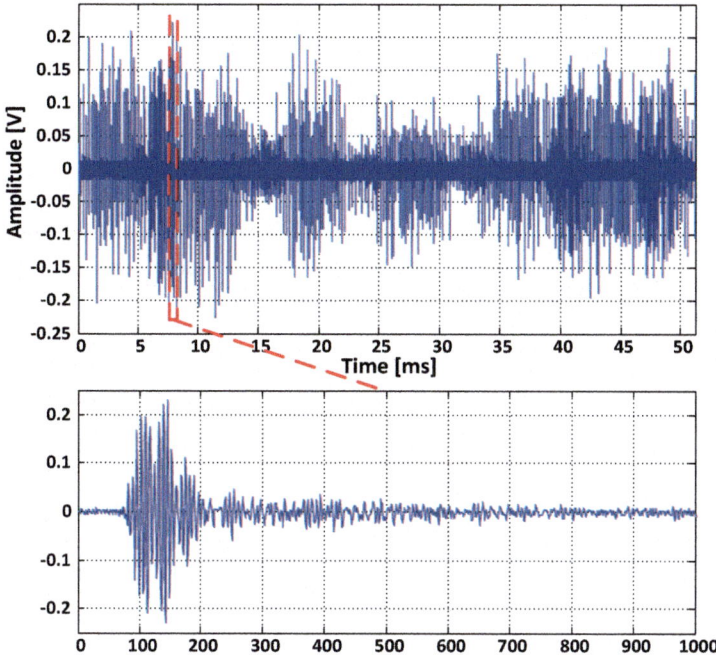

Fig. 6.3 Impulsive noise measured from substation 1, 315 kV area

an impulse occurs. An impulse is observable as soon as its amplitude is superior to the background noise envelope, so an impulse can be interpreted as noise when samples have a larger variance than the background noise, which also implies that the variance of the samples contains the variance of the background noise ($\sigma_i^2 = \sigma_0^2 + \delta$) as in Middleton models [24, 25, 91]. Our model is only based on observation, thus impulsive noise can be considered as a process switching from background noise to impulses and vice versa, and the variance of the samples corresponding to the impulses is superior to the background noise variance must be ensured. Secondly, based on observations from Fig. 6.3, the impulses can be assumed to occur independently from each other, which means that the occurrence times are i.i.d. According to the measurements, the impulses seem to occur with an IAT that follows an exponential distribution (Fig. 6.4). Moreover, as soon as an impulse occurs, the samples contained in the impulse are correlated to produce a damped oscillation waveform. Finally, assuming that there are three groups of impulses that gather the short, the average-length and the long impulses, which are respectively the small, the average and the large impulses in amplitude. The choice of three groups is justified based on observations from the measurements; it is noted that some impulses are very large in terms of amplitude, while others are close to the background noise level. In order to have a smooth representation of substation

Fig. 6.4 Histogram
normalized for the
Inter-arrival times in
substation 1, 735 kV area

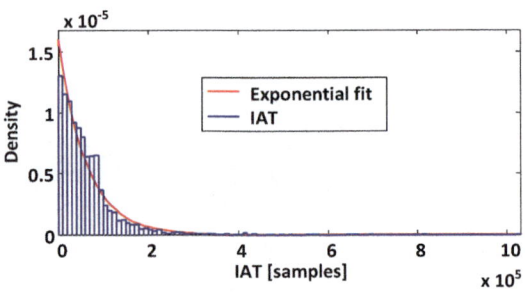

impulsive noise, a group of group medium impulses may be considered in favor of having impulses between the large impulses and the small impulses.

Parts of the proposed PMC model that contribute to the generation of correlated noise samples is introduced below:

- The background state corresponds to a Markov state associated with a zero-mean Gaussian distribution which represents the background noise;
- An impulsive state is a state of the partitioned Markov chain that does not represent the background noise and that is associated with a Gaussian distribution with a specific mean and variance;
- An impulsive system is a set of impulsive states connected with transitions probabilities in order to ensure a correlation between the samples;
- The impulse generation is configured with the transition probability p_{0i} from the background state to the system i;
- The impulse duration is configured with the probabilities to remain in the impulsive systems before returning the background state.

6.2   Impulsive System and Oscillations

The samples of each group of impulses are produces using a configuration of impulsive states, called "impulsive system" (Figs. 6.5 and 6.7), where the mean and the variance of the samples are estimated from the impulses of the groups observed.

The impulsive systems provide some oscillations within the impulse shape. The oscillations are implemented by using four or even six states in each impulsive system (Fig. 6.7) in order to generate samples around values characterizing a pseudo-sinusoidal signal. Such a waveform can be replicated by using Gaussian distributions with means and transitions carefully chosen; the samples should be organized with particular statistics in order to fit with a sinusoidal waveform as in Fig. 6.7.

In an impulsive system, a "loop" is the path of states when the process leaves an initial state, hits all other states in the system and returns to the initial state. For example, in Fig. 6.7, the paths of states $\{i_0, i_1, i_2, i_3, i_0\}$ and $\{i_0, i_1, i_2, i_3, i_4, i_5, i_0\}$ are "loops". The time spent in a loop corresponds to a

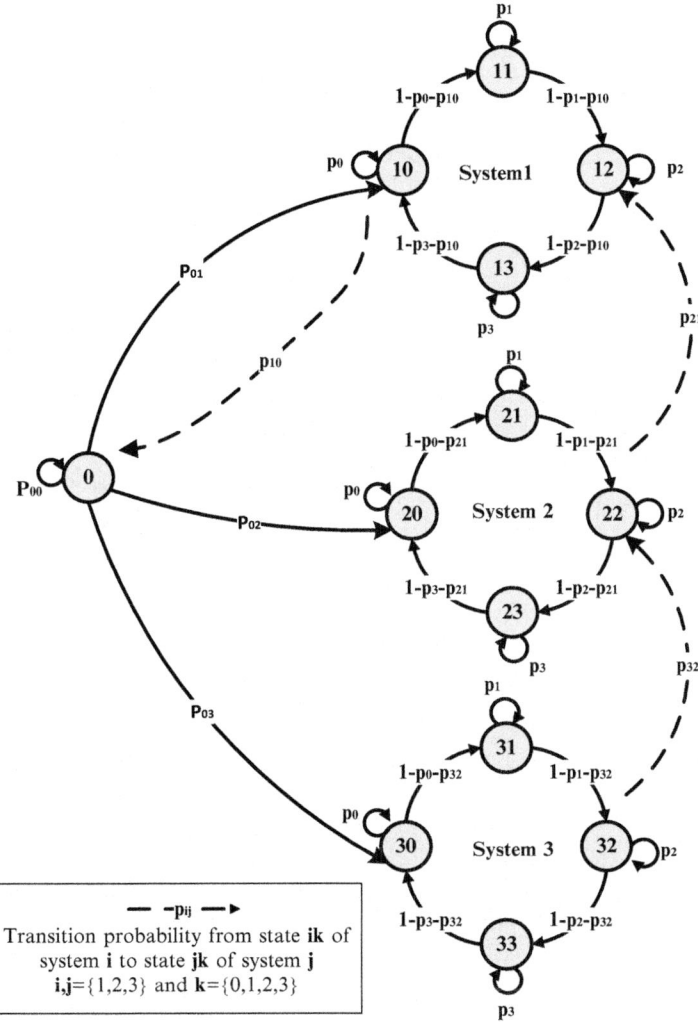

Fig. 6.5 Proposed model: partitioned Markov chain with systems of four states configured to generate the impulses samples with an oscillating waveform

sinusoidal period that have been evaluated from the oscillation frequency of the observed impulses. The number of states is inferior to the ratio of the sampling frequency to the desired frequency, because each state generates at least one sample; therefore the number of states is the minimum number of samples for one period of a sinusoid. For example, at a sampling frequency of 5 Giga-Samples per second, sinusoidal signals of frequency up to 1.25 GHz can be generated with four states. According to the spectrum of the measurements from substations (Fig. 6.2), the highest frequency peak is centered at 800 MHz, for which four or six states are enough.

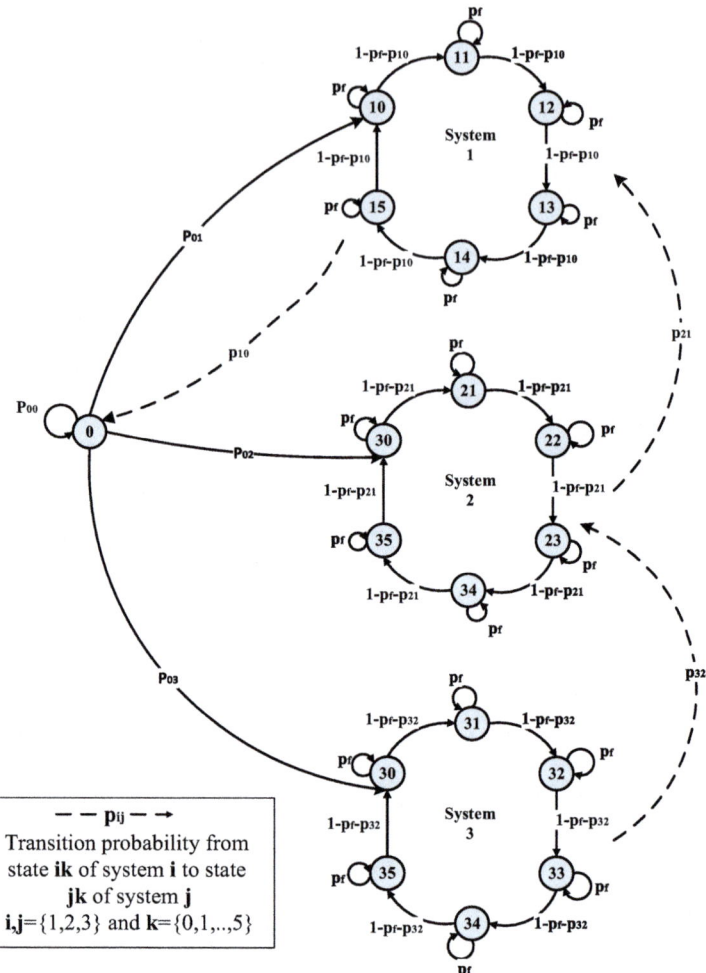

Fig. 6.6 Proposed model: partitioned Markov chain with systems of six states configured to generate the impulses samples with an oscillating waveform

A possible configuration is presented below.
For oscillations using four states (Fig. 6.5) :

- state i_0 : $\mathcal{N}(0, \sigma_0^2)$
- state i_1 : $\mathcal{N}(m_i, \sigma_i^2)$
- state i_2 : $\mathcal{N}(0, \sigma_0^2)$
- state i_3 : $\mathcal{N}(-m_i, \sigma_i^2)$

For oscillations using six states (Fig. 6.6):

- state i_0 : $\mathcal{N}(m_i/2, \sigma_0^2)$

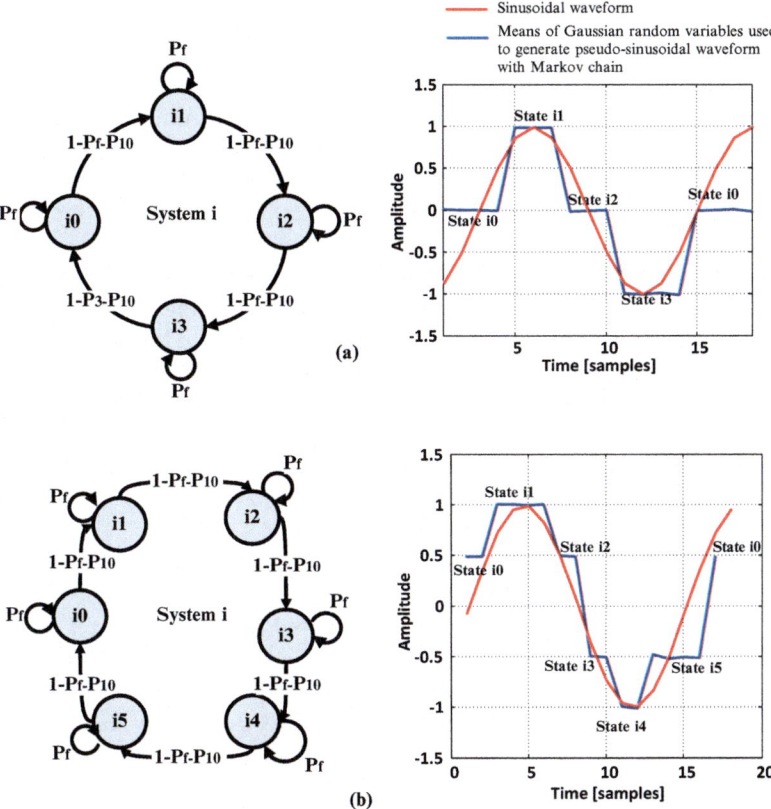

Fig. 6.7 Implementation of an oscillation with four states (**a**) and six states (**b**) in an impulsive system i

- state i_1 : $\mathcal{N}(m_i, \sigma_i^2)$
- state i_2 : $\mathcal{N}(m_i/2, \sigma_0^2)$
- state i_3 : $\mathcal{N}(-m_i/2, \sigma_0^2)$
- state i_4 : $\mathcal{N}(-m_i, \sigma_i^2)$
- state i_5 : $\mathcal{N}(-m_i/2, \sigma_0^2)$

where $\mathcal{N}(m, \sigma^2)$ represents the Gaussian distribution with mean m and variance σ^2. σ_0^2 and σ_i^2 are respectively the variances of the background noise and the impulse amplitude of the group i, and m_i is the mean of the impulse amplitude in group i.

This configuration of Gaussian distributions gives good results [140] but it can be improved by estimating the Gaussian parameter with a clustering algorithm [167]. Further details regarding this technique is discussed later in this chapter.

6.3   Damping Effect

In an impulsive system, each state is associated with a Gaussian distribution where
the mean and the variance of the samples are estimated from the impulses of the
groups observed. System i is the impulsive system that represents a impulse group
i having an average amplitude m_i. The damping effect is ensured by the succession
of systems with decreasing average amplitude m_i, $m_3 > m_2 > m_1$ (Fig. 6.5).
For example, the impulses in group 3 (larger amplitude and longer duration), are
generated by transition between states in system 3, then in system 2 and finally in
system 1. With the sum of the time spent in each system being equal to the duration
of the impulse in group i, the damping effect can be ensured with the appropriate
duration of the impulses.

6.4   Transition Matrix

The transition matrix parameterizes the events of impulsive noise such as the
impulse occurrence, the impulse duration and the oscillation. As explained, the
oscillations are performed with the "loop" in impulsive systems and the frequency
is set by selecting the appropriate probability to remain in each state. The diagonal
coefficients of the transition matrix T are all the same and they are estimated
from the largest frequency peak of the impulse spectrum. Only the first diagonal
coefficient is different from the others because it represents the probability to remain
in the background noise and it is estimated from the number of impulses that are
observed.

This matrix is designed to create appropriate time correlations between the
impulsive noise samples. Also it configures the occurrence times of the impulses
with the probabilities to go from the state 0 to the other systems i, $i = 1, 2, 3$
(Fig. 6.5). Basically, the transition matrix is set by using the information extracted
from the measurements. As soon as the three groups of impulses is detected (the
method is presented in Sect. 4.5.1), the mean duration of the impulses and the
oscillation frequency are analyzed. With the oscillation frequency, the probability
can be set in order to remain in each state of an impulsive system. To simplify the
parameter estimation, the same probability p_f to remain in each impulsive state is
used. Moreover, the time spent in a system is configured by setting the probability
$p_{i,i-1}$, $\{i = 1, 2, 3\}$ to go from the system i to the system $i-1$ (Fig. 6.5). Section 6.5
describes how to calculate the transition probabilities.

The transition matrix of the PMC model is configured as follows:

$$T = \begin{pmatrix}
P_{00} & P_{01} & P_{01} & P_{01} & P_{01} & P_{01} & P_{01} & P_{02} & P_{02} & P_{02} & P_{02} & P_{02} & P_{02} & P_{03} & P_{03} & P_{03} & P_{03} & P_{03} & P_{03} \\
P_{10} & P_f & q_{10} & 0 & 0 & 0 & 0 & 0 & 0 & 0 & 0 & 0 & 0 & 0 & 0 & 0 & 0 & 0 & 0 \\
P_{10} & 0 & P_f & q_{10} & 0 & 0 & 0 & 0 & 0 & 0 & 0 & 0 & 0 & 0 & 0 & 0 & 0 & 0 & 0 \\
P_{10} & 0 & 0 & P_f & q_{10} & 0 & 0 & 0 & 0 & 0 & 0 & 0 & 0 & 0 & 0 & 0 & 0 & 0 & 0 \\
P_{10} & 0 & 0 & 0 & P_f & q_{10} & 0 & 0 & 0 & 0 & 0 & 0 & 0 & 0 & 0 & 0 & 0 & 0 & 0 \\
P_{10} & 0 & 0 & 0 & 0 & P_f & q_{10} & 0 & 0 & 0 & 0 & 0 & 0 & 0 & 0 & 0 & 0 & 0 & 0 \\
P_{10} & q_{10} & 0 & 0 & 0 & 0 & P_f & 0 & 0 & 0 & 0 & 0 & 0 & 0 & 0 & 0 & 0 & 0 & 0 \\
0 & P_{21} & 0 & 0 & 0 & 0 & 0 & P_f & q_{21} & 0 & 0 & 0 & 0 & 0 & 0 & 0 & 0 & 0 & 0 \\
0 & 0 & P_{21} & 0 & 0 & 0 & 0 & 0 & P_f & q_{21} & 0 & 0 & 0 & 0 & 0 & 0 & 0 & 0 & 0 \\
0 & 0 & 0 & P_{21} & 0 & 0 & 0 & 0 & 0 & P_f & q_{21} & 0 & 0 & 0 & 0 & 0 & 0 & 0 & 0 \\
0 & 0 & 0 & 0 & P_{21} & 0 & 0 & 0 & 0 & 0 & P_f & q_{21} & 0 & 0 & 0 & 0 & 0 & 0 & 0 \\
0 & 0 & 0 & 0 & 0 & P_{21} & 0 & 0 & 0 & 0 & 0 & P_f & q_{21} & 0 & 0 & 0 & 0 & 0 & 0 \\
0 & 0 & 0 & 0 & 0 & 0 & P_{21} & q_{21} & 0 & 0 & 0 & 0 & P_f & 0 & 0 & 0 & 0 & 0 & 0 \\
0 & 0 & 0 & 0 & 0 & 0 & 0 & P_{32} & 0 & 0 & 0 & 0 & 0 & P_f & q_{32} & 0 & 0 & 0 & 0 \\
0 & 0 & 0 & 0 & 0 & 0 & 0 & 0 & P_{32} & 0 & 0 & 0 & 0 & 0 & P_f & q_{32} & 0 & 0 & 0 \\
0 & 0 & 0 & 0 & 0 & 0 & 0 & 0 & 0 & P_{32} & 0 & 0 & 0 & 0 & 0 & P_f & q_{32} & 0 & 0 \\
0 & 0 & 0 & 0 & 0 & 0 & 0 & 0 & 0 & 0 & P_{32} & 0 & 0 & 0 & 0 & 0 & P_f & q_{32} & 0 \\
0 & 0 & 0 & 0 & 0 & 0 & 0 & 0 & 0 & 0 & 0 & P_{32} & 0 & 0 & 0 & 0 & 0 & P_f & q_{32} \\
0 & 0 & 0 & 0 & 0 & 0 & 0 & 0 & 0 & 0 & 0 & P_{32} & q_{32} & 0 & 0 & 0 & 0 & 0 & P_f
\end{pmatrix},$$

with $q_{ij} = 1 - p_f - p_{ij}$ and $p_{00} = 6 \times p_{01} + 6 \times p_{02} + 6 \times p_{03}$.

The transitions from the background noise to the impulse groups are equal for each state of a group (p_{01}, p_{02} and p_{03}), which means that each state of a group has the same chance to generate the first sample of an impulse.

The maximum likelihood estimate of the transition probabilities is given by the following expression:

$$p_{ij} = \frac{\text{number of transitions from } i \text{ to } j}{\text{number of samples in state } i} \tag{6.1}$$

To configure the impulse duration, a first step is to find the average number of samples it takes for the Markov chain to leave an impulsive state. Focusing on state 0 for this example.

The probability to stay exactly n samples in state 0 after starting in state 0 is calculating as:

$$Pr\{V_0 = n\} = Pr\{X_n \neq 0, X_{n-1} = 0, \ldots, X_1 = 0 \mid X_0 = 0\}$$
$$= (1 - p_{00})p_{00}^{n-1}. \tag{6.2}$$

where V_0 is the time spent before leaving the state 0.

Equation (6.2) provides the distribution of the number of samples spent in state 0, and now, the average time spent in state 0 can be calculated.

Applying the following equation to calculate the mean time spent in state 0:

$$E[V_0] = \frac{1}{1 - p_0} \tag{6.3}$$

The time spent in a state is configured by the probability to leave it, therefore the time spent in an impulsive system can be configured in the same way by using the probability to leave a system (i.e. p_{32}, p_{21} and p_{10} in Fig. 6.5). If we consider an impulsive system as a single state, the probability to leave the system is the transition probability from any state of the system to another system or to the background noise state. To configure an average time T_i, in samples, spent in an impulsive system, the probability to leave each state of the impulsive system to another system, or to the background noise, is $p = 1 - \frac{1}{T_i}$. T_i is estimated from the sample mean of the impulse duration of a group.

To configure the oscillation frequency within the impulses, the model must have the appropriate probabilities to remain in the impulsive states in each system; with the probability to remain in an impulsive state, the model ensures an average sample duration for each state, which also contributes to ensure an average period of a sinusoid signal in an impulsive system (Fig. 6.7). Intuitively, the average number of samples spent in a period (T_{osc}) of an impulsive system seems to be the sum of the average times spent in each state ($T_{osc} = \sum_i \frac{1}{1-p_i}$). To verify this assumption, the average number of samples, T_{osc} to return to a state after leaving it, in an impulsive system is calculated.

Choosing the example of a system of four states (Fig. 6.7) to illustrate the concept; the average time that we are interested in is the number of samples it takes for the process to return to the state $i0$ after leaving it. To study the average time of a "loop", we have to get rid of the other possible paths that might come after the transition from state $i3$ to state $i0$ and that do not characterize an oscillation period, therefore we modify the impulsive system in Fig. 6.7 by adding an absorbing state accessible after leaving the state $i3$ (Fig. 6.8). The absorbing state (state $i4$ in Fig. 6.8) represents any possible state that might come after returning to state $i0$. The average time of a loop, which is also the average period of the quasi-sinusoidal signal that the impulsive system attempts to replicate, is the average time before absorption when the initial state is $i0$. We calculate the average time to absorption [168] by using the transition matrix of the Markov chain of an impulsive system. For an absorbing Markov chain, the transition matrix has the following form:

$$P = \left(\begin{array}{c|c} Q & \vdots \\ \hline \cdots & I \end{array} \right)$$

Fig. 6.8 Modified impulsive system with an absorbing state $i4$

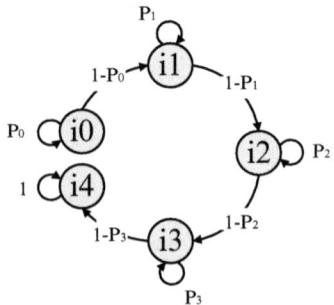

where Q is the submatrix representing the process as long as it remains in transient states and I is the identity matrix representing the absorbing state [168]. The transition matrix is defined as:

$$P = \left(\begin{array}{cccc|c} p_0 & 1-p_0 & 0 & 0 & 0 \\ 0 & p_1 & 1-p_1 & 0 & 0 \\ 0 & 0 & p_2 & 1-p_2 & 0 \\ 0 & 0 & 0 & p_3 & 1-p_3 \\ \hline 0 & 0 & 0 & 0 & 1 \end{array}\right)$$

$$Q = \left(\begin{array}{cccc} p_0 & 1-p_0 & 0 & 0 \\ 0 & p_1 & 1-p_1 & 0 \\ 0 & 0 & p_2 & 1-p_2 \\ 0 & 0 & 0 & p_3 \end{array}\right)$$

the fundamental matrix: $N = (I - Q)^{-1}$ is introduced [168]

$$N = \left(\begin{array}{cccc} \frac{1}{1-p_0} & \frac{1}{1-p_1} & \frac{1}{1-p_2} & \frac{1}{1-p_3} \\ 0 & \frac{1}{1-p_1} & \frac{1}{1-p_2} & \frac{1}{1-p_3} \\ 0 & 0 & \frac{1}{1-p_2} & \frac{1}{1-p_3} \\ 0 & 0 & 0 & \frac{1}{1-p_3} \end{array}\right)$$

According to [168], the average time before absorption from any of the non-absorbing states can be found using the following vector t, which is the vector of the average times before absorption after leaving each state.

$$t = \left(\begin{array}{cccc} \frac{1}{1-p_0} & \frac{1}{1-p_1} & \frac{1}{1-p_2} & \frac{1}{1-p_3} \\ 0 & \frac{1}{1-p_1} & \frac{1}{1-p_2} & \frac{1}{1-p_3} \\ 0 & 0 & \frac{1}{1-p_2} & \frac{1}{1-p_3} \\ 0 & 0 & 0 & \frac{1}{1-p_3} \end{array}\right) \times \left(\begin{array}{c} 1 \\ 1 \\ 1 \\ 1 \end{array}\right)$$

$$= \left(\begin{array}{c} \frac{1}{1-p_0} + \frac{1}{1-p_1} + \frac{1}{1-p_2} + \frac{1}{1-p_3} \\ \frac{1}{1-p_1} + \frac{1}{1-p_2} + \frac{1}{1-p_3} \\ \frac{1}{1-p_2} + \frac{1}{1-p_3} \\ \frac{1}{1-p_3} \end{array}\right)$$

$$(6.4)$$

We are interested in the time spent in the Markov chain before absorption when the initial state is 0, which corresponds to the first coefficient of vector t. The average time to absorption is $\frac{1}{1-p_0} + \frac{1}{1-p_1} + \frac{1}{1-p_2} + \frac{1}{1-p_3}$, which is the sum of the mean times spent in each non-absorbing state.

To lower the complexity of the model, each pair of states in the impulsive systems has the same transition probability. For a system using six states, the average time spent in a loop is $\frac{6}{1-p}$. For a given impulse frequency f in $samples^{-1}$, each state in an impulsive system must have a probability $p_i = 1 - 6f$. For the simulations, the impulsive systems with six states are considered because this configuration offers a better implementation of the oscillations.

6.5   Parameter Estimation

The partitioned Markov chain requires parameters for the time characteristics and the impulse waveform. The parameters associated with time characteristics of the impulses are the transitions probabilities of the Markov chain. There are three kinds of time parameters:

- The impulse duration, which is configured with the probabilities to remain in the impulsive systems;
- The oscillation period, which is configured with the probability to remain in the states of a system. The probability is calculated form the highest frequency peak of the impulses spectrum;
- The impulse occurrence, which is determined by the transition from the state 0 to the impulsive systems.

The parameter estimation for the PMC model requires to work on two aspects of the noise: the sample value and the time characteristics (impulse duration and repetition rate). The impulses are detected based on the sample value above the background noise threshold th_a. From Fig. 6.9, it is noted that a possible use of the correlation between the impulse duration and the amplitude for the definition of group of impulse. The PMC model is configured to represent three groups of impulses by assuming that an impulse with large amplitude lasts longer. The clusters 1, 2 and 3, as shown in Fig. 6.9, represent respectively the short, the average and the large impulses that are gathered in groups 1, 2 and 3.

6.5.1   Fuzzy C-Means Algorithm

In order to estimate the required parameters for the clusters in Fig. 6.9, a fuzzy C-means is used, and it is described below.

The fuzzy C-means algorithm (FCM) classifies a sequence of data $X = [x_1, x_2, \ldots, x_N]$ into M clusters with centers $C = [c_1, c_2, \ldots c_M]$, $N > M$.

The cluster centers and radii are found by minimizing the following objective function:

$$J(X; \Phi, C) = \sum_{i=1}^{M} \sum_{k=1}^{N} (\phi_{ik})^m d^2(x_k, c_i) \qquad (6.5)$$

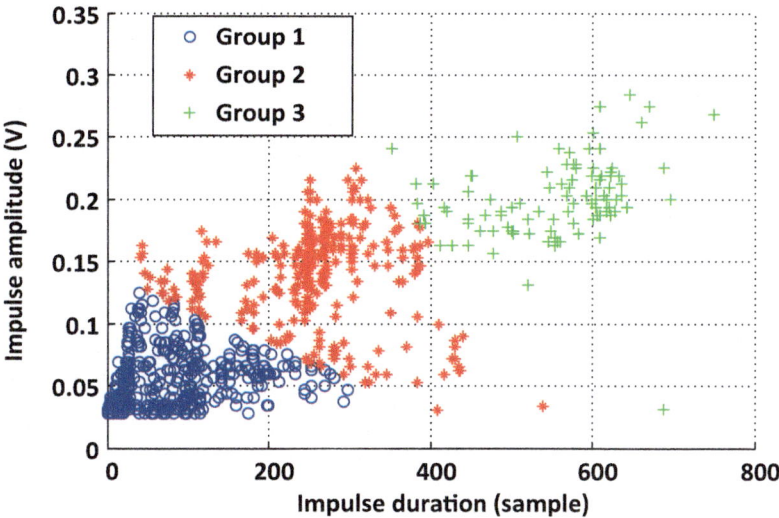

Fig. 6.9 Clusters for parameter estimation

with $\Phi = [\phi_{ik}]$ the membership sequence between a point k and a cluster i, such that $\sum_{i=1}^{M} \phi_{ik} = 1$ and $\forall i,\ 0 < \sum_{i=1}^{N} \phi_{ik} < N$ and $m = 2$.

This algorithm is used for two purposes:

- Classify the impulses into three groups given the correlation between their amplitude and duration. In this case, X in (6.5) is made up of x_k that have as coordinates the kth impulse duration and the kth impulse amplitude, for $k = 1, 2, \ldots, N$. A two dimensional plane is drawn, in which each point represent a single impulse with the amplitude and the time as coordinates (Fig. 6.9).
- For each group, find the Gaussian parameters that will be used to generate the sample value for each Markov state. In this case, in (6.5) $x_k = k^{th}$ sample value, $k = 1, 2, \ldots N$.

To estimate the Gaussian distribution parameters that are associated with each Markov state, three signals composed of impulse samples, where each impulse is identified as belonging to a specific group $i,\ i = \{1, 2, 3\}$ are created. These three signals are described as follows:

- $Signal_1$ is a signal composed of all samples in impulses from group 1; This corresponds to the small impulses.
- $Signal_2$ is composed of the samples within the impulses of group 2 that form "the stronger part of the impulse". The stronger part is shown in Fig. 6.10 and it corresponds to the samples of the impulses from the first sample to two times the rise time T_r. In this case, a power spectrum of PMC impulse more similar to the power spectrum calculated from the measurements is expected. The rise time

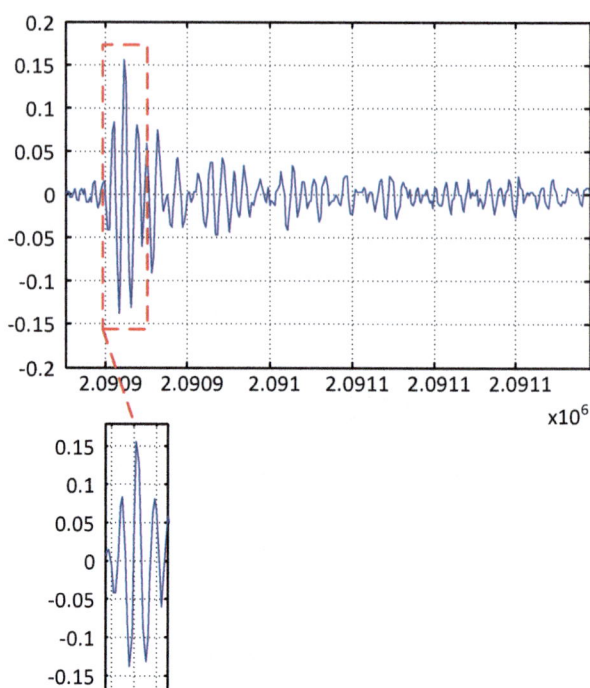

Fig. 6.10 Stronger part of an impulse

is defined as the duration of an impulse between the first sample and the sample that is identified as the amplitude of the impulse.

- $Signal_3$ is formed exactly with the same method as $Signal_2$, except the samples used are from the impulses in group 3.

When the three signals are formed, the FCM algorithm is used, which is configured with four clusters using the samples of each signal. An example of the samples classified for $Signal_2$ is showed in Fig. 6.11 using the FCM algorithm. Each sample of the signal can be identified, and a graphic label is assigned to indicate if the sample belongs to cluster 1, 2, 3 or 4. The samples are classified into cluster and the algorithm provides the center and the radius that will be used to parameterize the Gaussian distributions for each state of the PMC model.

6.6 Results

Using the transition matrix configured with probabilities that are estimated from the measurements, a sequence of 256 millions of samples is generated. The PMC model manages to replicate the correlation between the samples in order to shape the impulse with a damped oscillating waveform (Fig. 6.12). It is noted that the waveform from the generated impulses is more similar to the measurements than

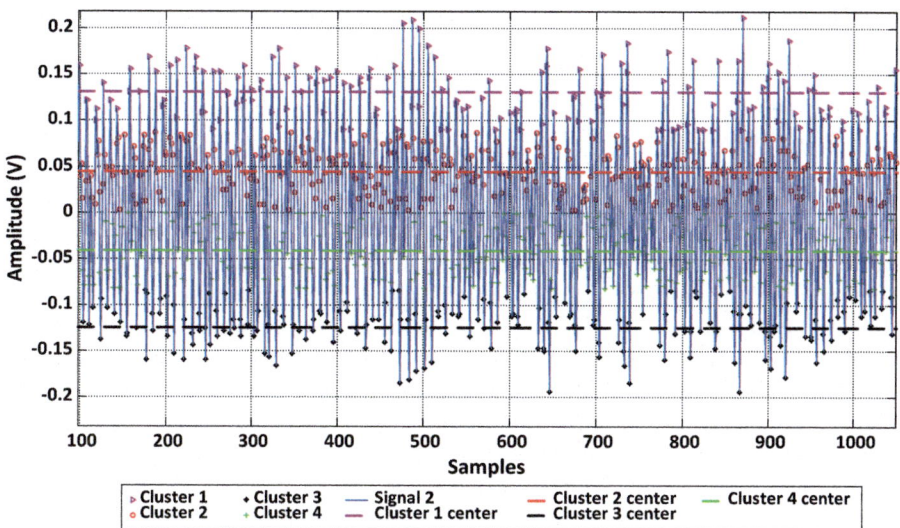

Fig. 6.11 Cluster for the Gaussian distribution associated with the states of the Markov chain: example of $Signal_2$

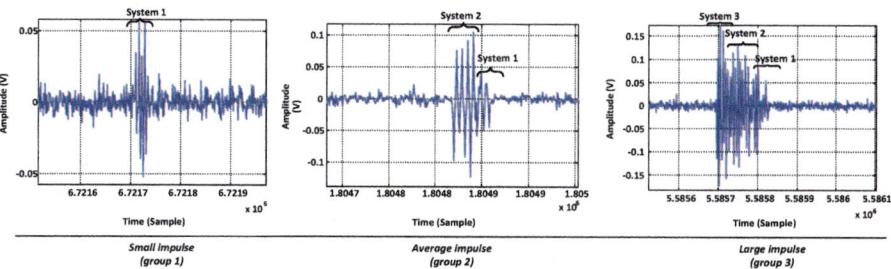

Fig. 6.12 Impulses from PMC model

the impulses that are generated with MC2 model. A statistical study of the distance between the measurements and the noise generated by existing models and the proposed PMC is conducted.

6.6.1   Divergence Between Measurements and Models

The PMC model with impulsive noise sequences collected from substations for different voltage ranges (230, 315 and 735 kV) is tested. In order to evaluate the performances of our model, we study the divergence between the distribution of the measurements and our model for the impulsive noise characteristics. The Kullback-Leibler divergence (KL-divergence) coefficient is calculated, which is a non-symmetric measure to estimate how close a distribution is to another [78]. The KL-divergence coefficient over a set X is given by : $coeff_{KL} = \sum_{x \in X} log(\frac{p(x)}{q(x)}) p(x)$, where p and q are respectively the distribution of the impulsive noise characteristic coming from the measurements and the model. The Mean Square Error (MSE) of the cumulative distribution function (CDF) between the distributions coming from the proposed model with the distributions coming from the measurements is calculated. For the same impulsive noise characteristics, the divergence coefficients of the measurements from the proposed PMC model but also from the Bernoulli-Gaussian with memory and the Middleton class-A models are compared.

For the three models, 256 millions samples are generated, and the noise sequences are analyzed to extract the distributions of the impulsive noise characteristics. The Middleton PDF is used to generate i.i.d. samples, then it only helps to compare the distribution of the noise samples' value within the impulses. The impulse samples are collected using the impulse detection method for the Bernoulli-Gaussian with memory and for our model. For Middleton model, the impulse samples are directly collected while generating the noise sequence using (2.15): a sequence is composed of samples provided by the Gaussian distributions representing the impulses, $m > 0$ in (2.15).

According to Tables 6.1, 6.2 and 6.3, we observe that our model performs better than the Bernoulli-Gaussian with memory for the distributions of the impulses characteristics and better than the other models for the distribution of the samples' value within the impulses.

The divergence tables indicate that the proposed PMC model generates impulses with a duration distributed more similarly to the measurements than the Bernoulli-Gaussian model. The PMC model considers three groups of impulses while the Bernoulli-Gaussian model considers just one; hence more precise and a more accurate distribution for the impulse duration are provided.

Table 6.1 Divergence for the impulsive noise characteristics between the measurements and the impulsive noise models; 230 kV substation

	Class-A	BG with memory	PMC model
KL-divergence			
Sample value	0.8585	0.090	**0.0549**
Impulse duration	–	0.6149	**0.4325**
IAT	–	0.7582	**0.6901**
Impulse amplitude	–	1.2102	**0.7579**
MSE of CDF			
Sample value	1.5897	0.405	**0.0446**
Impulse duration	–	2.1856	**0.8563**
IAT	–	0.8722	**0.6458**
Impulse amplitude	–	4.7873	**0.7281**

(best results are written in bold)

Table 6.2 Divergence for the impulsive noise characteristics between the measurements and the impulsive noise models; 315 kV substation

	Class-A	BG with memory	PMC model
KL-divergence			
Sample value	0.5966	0.1793	**0.1587**
Impulse duration	–	0.1466	**0.0877**
IAT	–	0.1422	**0.0909**
Impulse amplitude	–	0.3195	**0.2005**
MSE of CDF			
Sample value	0.5518	0.1209	**0.1070**
Impulse duration	–	0.275	**0.0109**
IAT	–	0.1122	**0.0439**
Impulse amplitude	–	0.3397	**0.0679**

(best results are written in bold)

Table 6.3 Divergence for the impulsive noise characteristics between the measurements and the impulsive noise models; 735 kV substation

	Class-A	BG with memory	PMC model
KL-divergence			
Sample value	0.8408	0.3339	**0.1263**
Impulse duration	–	0.2486	**0.2382**
IAT	–	0.5542	**0.5219**
Impulse amplitude	–	0.5146	**0.3221**
MSE of CDF			
Sample value	1.4016	0.2768	**0.0983**
Impulse duration	–	0.0883	**0.0798**
IAT	–	0.2450	**0.2099**
Impulse amplitude	–	0.1995	**0.1027**

(best results are written in bold)

The IAT seems to be distributed equivalently for the PMC model and the Bernoulli-Gaussian model with memory; the divergence coefficients are not so different between the models, which makes sense because the impulse occurrence is basically implemented in the same way for both models. However, our model offers better results because the process that generates the impulse occurrence is dependent of the process that generates the impulse duration. The partitioned Markov chain can only generate the first sample of an impulse while the current state is the background noise, then a more accurate implementation of the impulse duration will also provide a more accurate implementation of the IAT.

The proposed PMC model also generates impulse sample value with the best similarity with the measurements compared to the Middleton class-A and the Bernoulli-Gaussian with memory models. This good performance can be explained by the use of 18 Gaussian distributions to generate the impulse samples. With 18 Gaussian distributions estimated with the proposed method and with the configu-

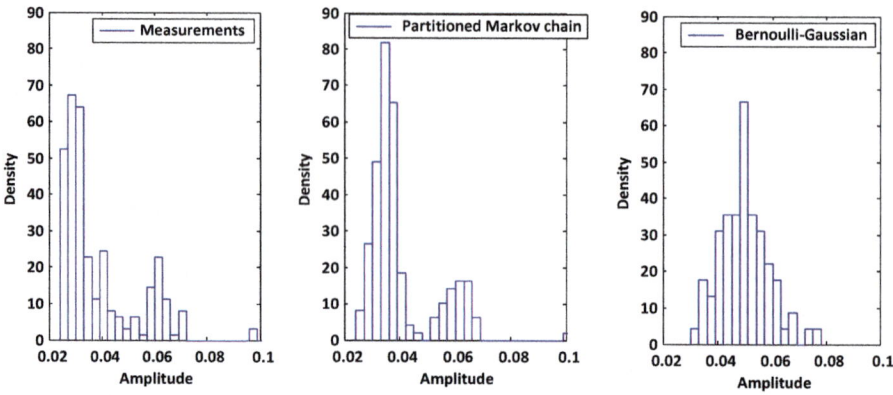

Fig. 6.13 Amplitude of the impulses extracted from the models and the measurements

ration of transitions between the Markov states, the PMC model manages to create impulse waveforms very similar to the measured impulses.

The three groups of impulses also give good results for the distribution of the impulse amplitude for our model. For the KL divergence and the MSE of the CDFs we have a more significant difference between our model and the Bernoulli-Gaussian. This difference can be explained by the implementation of the impulsive systems that uses Gaussian distributions with a non-zero mean and a variance, while the Bernoulli-Gaussian model only uses a zero-mean Gaussian distribution with a larger variance. This difference can be observed on the histograms of the impulse amplitude for both models and the measurements (Fig. 6.13). The distribution of the three groups of amplitude observed from the measurements and from the PMC model can be distinguished, which cannot provide the Bernoulli-Gaussian model.

6.6.2 Spectrum Analysis

Another characteristic that is important for a wide-band representation of the noise is the power spectrum. All the impulses from the measurements are gathered to calculate the average power spectrum and we did the same for the noise generated by the proposed PMC model (Fig. 6.14). Focusing on classic wireless frequencies, respectively 900 MHz, 1.8 GHz, and 2.4 GHz, our model is, on average, 3 dB from the measurements. The other models cannot provide such a power spectrum, because they do not provide the appropriate correlation between the samples within the impulses. According to our last experiment in Sect. 4.6.1 with the PMC model, all the impulsive noise characteristics are better replicated than using the other classic models (i.e., Middleton class-A and MC2). The main contribution of the PMC model remains the power spectrum of the impulses that replicates the spectrum measured while the power spectrum of the other models (Middleton class-A and MC2) provide impulses with a flat spectrum (Fig. 6.14).

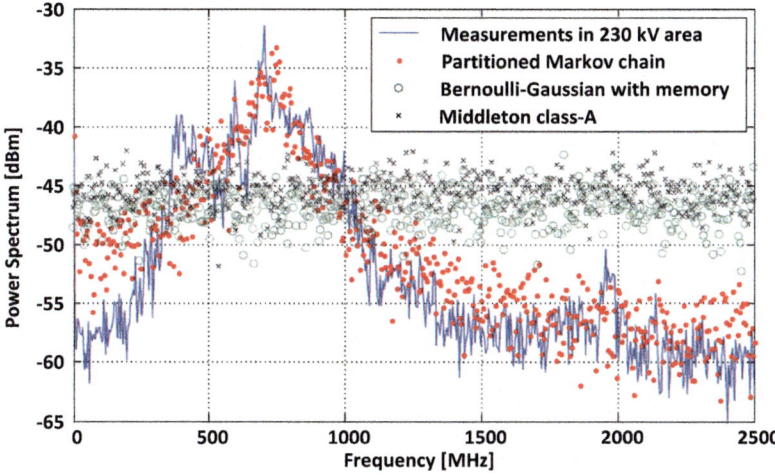

Fig. 6.14 Power spectrum of noise generated by proposed system as compared to measurements

6.7   Representative Parameters for PMC Model in Wide Band

Each model requires parameters that can be estimated using noise samples; also each parameter has a meaning in terms of impulsive noise characteristics, which means that they can be calculated using their definition and the information extracted from the noise characteristics.

In order to estimate representative parameters, all the recorded noises are used. First impulse detection and classification for one voltage of equipment must be performed. These allow the extraction of information that is required by models using the FCM algorithm (Fig. 6.15). The procedure is proceeded as follows:

- Create $signal_{BN}$, composed of a sequence of one millions of background noise sample from each measurement file
- Create $signal_1$, $signal_2$ and $signal_3$, composed as in Sect. 4.5.1;
- Classify the impulse duration and rise times;
- Create $signal_0$, composed using all the impulse samples;
- Classify the number of impulses for each group to provide the repetition rate.

The parameters for the PMC model are estimated with the method explained in Fig. 6.15 (Tables 6.4 and 6.5).

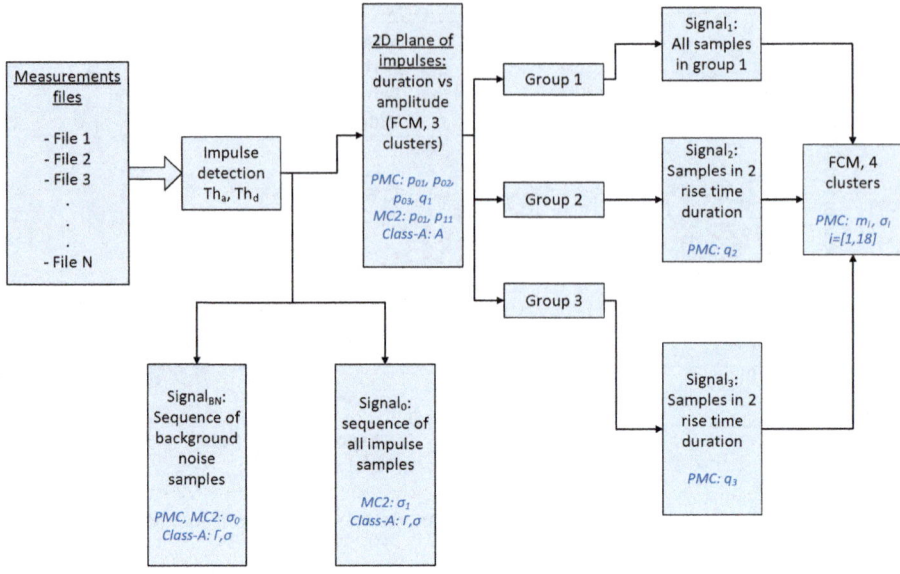

Fig. 6.15 New signal building for parameter estimation

Table 6.4 PMC parameters: transition matrix

PMC parameters	Voltage area			
	25 kV	230 kV	315 kV	735 kV
p_{01}	4.221×10^{-8}	2.315×10^{-7}	3.730×10^{-7}	3.624×10^{-7}
p_{02}	1.614×10^{-8}	1.149×10^{-7}	1.681×10^{-7}	1.507×10^{-7}
p_{03}	9.609×10^{-9}	7.952×10^{-8}	8.489×10^{-8}	3.203×10^{-8}
p_{10}	0.027	0.044	0.031	0.036
p_{21}	0.018	0.028	0.022	0.058
p_{32}	0.013	0.027	0.012	0.05
p_f	0.064	0.064	0.064	0.064

6.8 Conclusions

An impulsive noise model in wide band (780 MHz to 2.5 GHz) using a partitioned
Markov chain configuration has been designed. The proposed PMC model generates
correlated samples, which produces an impulsive noise very similar to the noise
measured in substations. Although the model requires more distributions than other
models, the results are very satisfying since the impulses generated with PMC
are more similar to the impulses measured in substations in terms of impulsive
noise characteristics, such as impulse duration, impulse amplitude, IAT and power
spectrum of the impulses.

Representative parameters for the proposed PMC model have been calculated,
which will allow anyone to represent an RF environment of substations, based on the

Table 6.5 PMC parameters: Gaussian distributions

| | Voltage area | | | | | | | |
| | 25 kV | | 230 kV | | 315 kV | | 735 kV | |
State	m_i	σ_i	m_i	σ_i	m_i	σ_i	m_i	σ_i
0	0	0.0019985	0	0.0045068	0	0.0051904	0	0.0058157667
1	0.0042472	0.0025758	0.0086062	0.0056734	0.012456	0.0082745	0.014263151	0.011258112
2	0.013584	0.0032054	0.030955	0.0074989	0.046107	0.015	0.057280523	0.021613978
3	0.0042472	0.0025758	0.0086062	0.0056734	0.012456	0.0082745	0.014263151	0.011258112
4	-0.0049313	0.0025516	-0.010185	0.00597	-0.014415	0.0084192	-0.021453075	0.011065931
5	-0.014455	0.0032753	-0.032708	0.0072949	-0.047658	0.015269	-0.063524956	0.021508378
6	-0.0049313	0.0025516	-0.010185	0.00597	-0.014415	0.0084192	-0.021453075	0.011065931
7	0.0057005	0.0037871	0.010311	0.0072848	0.017074	0.014179	0.021440885	0.02816402
8	0.021085	0.0082377	0.041409	0.013753	0.088877	0.042745	0.15216049	0.049925421
9	0.0057005	0.0037871	0.010311	0.0072848	0.017074	0.014179	0.021440885	0.02816402
10	-0.0070886	0.0038219	-0.012454	0.0074523	-0.021611	0.014888	-0.058457989	0.029384957
11	-0.023008	0.0083686	-0.043204	0.013022	-0.09125	0.043309	-0.18588027	0.053559871
12	-0.0070886	0.0038219	-0.012454	0.0074523	-0.021611	0.014888	-0.058457989	0.029384957
13	0.0074193	0.005131	0.017206	0.013331	0.028446	0.039275	0.022306216	0.060278796
14	0.029224	0.01305	0.067859	0.022572	0.31293	0.11782	0.35444699	0.11031369
15	0.0074193	0.005131	0.017206	0.013331	0.028446	0.039275	0.022306216	0.060278796
16	-0.0089195	0.0051435	-0.026773	0.01327	-0.05713	0.043372	-0.16732094	0.067299815
17	-0.030518	0.012076	-0.075642	0.021935	-0.36505	0.11959	-0.42970418	0.098973007
18	-0.0089195	0.0051435	-0.026773	0.01327	-0.05713	0.043372	-0.16732094	0.067299815

voltage of the equipment, without performing any measurement. The PMC model is in wide band, which means it is independent from the wireless technology used, but it is possible to filter the output of the model in order to have a narrow-band noise.

Chapter 7
Impulsive Noise in a Poisson Field of Interferers in Substations

7.1 Introduction

In Chaps. 4–6 accurate impulsive noise models have been proposed. These can be used for performance analyses of robust wireless communication systems as well as the development of diagnostic tools in high voltage equipments. The development of rapid and online methods of detection, identification and/or localization of partial discharge (PD) sources using wireless intelligent electronic devices is an area of growing interest [169–172]. The design of more accurate, rapid and efficient signal processing algorithms can be achieved with tractable and reasonable assumptions underlying the physics of such radio frequency (RF) signals/noise in substations. This chapter deals with statistical characterization and analysis of impulsive noise in situations in which RF impulsive signals are caused by electromagnetic radiations from PD sources.

In this chapter, a statistical analysis of transient impulsive noise in substation environments is conducted. A new and generalized impulsive noise model will be implemented using the Poisson field of interferers, from which first- and second-order statistics can be derived analytically based on the physics of the noise process. The statistical analysis allows for the identification of some interesting statistical properties of moments, cumulants and probability distributions. These can, in turn, be utilized in signal processing algorithms for rapid PD identification, localization, and impulsive noise mitigation techniques in wireless communications in substations.

The chapter is structured as follows: in Sect. 7.2, a mathematical formulation of multiple PD interference sources is presented, based on reasonable assumptions regarding the physical process of PD and the propagation of EM waves. By assuming that PD sources are randomly distributed over space-time, the Poisson-field of interferers allows for the generalization of an impulsive noise model in the presence of multiple PD sources in substations. In Sect. 7.3, probability

© Springer International Publishing AG, part of Springer Nature 2019
B. L. Agba et al., *Wireless Communications for Power Substations: RF Characterization and Modeling*, Wireless Networks,
https://doi.org/10.1007/978-3-319-91328-5_7

distributions and first- and higher-order moments of such noise processes are derived by taking advantage of the Poisson field of interferers. In Sect. 7.4, experimental and simulation results are presented to assess the goodness-of-fit of the proposed filtered Poisson process and existing impulsive noise models used by [45, 71, 102] and [72] for substation environments. An estimation procedure is proposed in order to compare first- and second-order statistics. Results show that it is more appropriate to use the filtered Poisson process rather than memoryless impulsive noise models when the impulsive component has transient effects. Section 7.5.1 is an application of the generalized model, by which a new technique for a rapid and online identification of PD sources is proposed using a blind-source separation technique. An estimation of a number of PD sources can help to evaluate the insulation performance and lifespan of HV equipment.

7.2 A Mathematical Formulation of Multiple PD Interference Sources

In this section, we define a mathematical formulation of a noise process in the presence of multiple PD interference sources based on tractable and reasonable physical assumptions, thereby establishing a generalized impulsive noise model for substation environments.

7.2.1 Electromagnetic Radiations of Multiple PD Sources

Assuming that a given receiver is surrounded by an arbitrary number of HV equipment under normal operation in a spatial region, $\Upsilon \in \mathbb{R}^d$ where d is the dimension of the space. For simplicity, the spatial region is restricted to a sphere in the three-dimensional Euclidean space where the receiver is located at the origin.

7.2.1.1 The Emission of the PD Impulses

Inside this spatial region Υ, each HV installation generates an arbitrary number of PD sources in which the induced electromagnetic radiations produce transient impulsive waveforms. PD impulses are characterized by a rapid impulse whose durations, rise-time, fall-time and amplitude are linked to the physical characteristics of PD, such as over-voltages, ionization processes, and/or free-electron rates [55, 173]. In such circumstances, it is reasonable to assume that all of the PD sources share a common random mechanism in which the original transient impulses induced by PD sources have the same type of waveform. This is denoted by $aD(\bar{\theta}, t)$, where a is a real positive random amplitude, and $\bar{\theta}$ is a set of time-invariant random variables denoting the duration, rise-time, and spectral characteristics.

7.2.1.2   Basic Assumptions of Spatial and Temporal PD Events

Inside the spatial region Υ, we assume that the number of HV equipment is countable when their positions are randomly distributed over space. Moreover, each HV installation has an arbitrary and countable number of PD sources where their positions are denoted by $\mathbf{r}_1, \mathbf{r}_2, \cdots$ and $\mathbf{r}_k \in \Upsilon$ is a random variable. As a result, the distance between the kth PD source and the receiver is written as $r_k = \|\mathbf{r}_k\|$, since the receiver is located at the origin of the spatial region.

In Chaps. 3 and 4, we described that the emission of a single PD is linked to both the local electric field and the number of free-electrons, which depend on the physical characteristics of PD sites as well as the applied voltage, thermal aging, and electrical insulation of an HV installation. Under such conditions, the amplitude and the time occurrence of the PD can be described as time-dependent random variables. In particular, the time occurrence follows a cyclostationary process due to the AC voltages. However, since the three-phase AC voltage is used in substations, the cyclostationary condition might not hold due to the superposition of all impulses from multiple PD sources. As a result, it is reasonable to assume that an impulse $a_k D(\bar{\theta}_k, t)$ occurs independently at random time $t_1, t_2, \cdots \in \mathbb{R}^+$. In addition, we shall assume that the random amplitude a_k and the random parameter characterizing the impulse $\bar{\theta}_k$ are independent and identically distributed (*iid*) whose statistical distributions are respectively given by $p_a(a)$ and $p_{\bar{\theta}}(\bar{\theta})$. PD sources are also assumed to be independent; in other words, the location $\mathbf{r}_k$, the temporal event of the impulse t_k, the random amplitude a_k and the random parameter $\bar{\theta}_k$ of the kth source are independent for all PD sources $k = 1, 2 \cdots$.

The resulting noise process observed by the receiver is a superposition of all transient PD impulses activated in the spatial region Υ. Since the receiver and PD sources are located in distinct positions, a transient impulse will be attenuated and distorted by the propagation channel and the receiver. With reasonable assumptions of the propagation conditions of EM waves, and the spatial and temporal distribution of PD sources, the statistical properties of the received noise process can be derived.

7.2.2   Propagation of EM Waves Induced by PD Sources

Statistical properties of the received noise process can be derived with suitable assumptions of the basic impulsive waveform of interferences sources. Middleton has specified a very general waveform when the physical process of source emission is taken into consideration for major classes of noise processes in [174, 175]. Unlike his expression, which requires some restrictions on the bandwidth of the receiver, we may simplify the expression by assuming that all of the PD sources inside the spatial region have the same isotropic radiation pattern and the receiver has an omni-directional antenna. This assumption is applicable for both narrowband and wideband noise processes.

7.2.2.1 The Noise Process Observed by the Receiver

An emitted PD impulse observed by the receiver is distorted and attenuated by both a propagation channel and the receiver itself. These distortions can be determined by the beam patterns of the PD source and the antenna, source location, and the impulse response of the receiver, including RF and IF stages of linear filters. As conjectured in Chap. 5, a PD transient impulse might be attenuated and distorted by the multipath effects induced by the presence of multiple reflected EM waves of the emitted impulse and the impulse response of the receiver by which the resulting impulsive signal has a transient effect. In such instances, the distortion is produced by the convolution product of an original transient PD impulse $aD(\bar{\theta}, t)$ located at $\mathbf{r}$ and the impulse response of the receiver, such that:

$$
\begin{aligned}
u_k(\theta, t, \mathbf{r}) &= c_k(\mathbf{r})aD_k(\bar{\theta}, t) * h_k(t) \\
&= c_k(\mathbf{r})u_k(\theta, t)
\end{aligned}
\tag{7.1}
$$

where $h(t)$ is the impulse response of both the propagation channel and the receiver. $c(\mathbf{r})$ is the attenuation factor function related to the distance between the PD source and the receiver. This can be determined by the beam patterns of sources and the receiver. θ is a set of random variables characterizing the distorted PD impulse observed by the receiver after any RF and IF stages of linear filtering.

In the presence of multiple PD sources, the resulting noise process is given by the superposition of all PD impulses, such that the noise is observed by the receiver is written as:

$$
X(t) = \sum_{k=1}^{N_I} u_k(\theta_k, t, \mathbf{r}_k)
\tag{7.2}
$$

where N_I is the total number of PD impulses arriving at the receiver within a given time observation. This will be defined from assumptions with respect to the spatial and temporal distribution of the PD sources. Next, the definition and emulation of the attenuated and distorted transient impulse $u(\theta_k, t_k, \mathbf{r}_k)$ are examined in detail.

7.2.2.2 A Generic Temporal Impulsive Waveform from PD

According to measurements in substations and laboratories, a generic transient impulsive waveform from PD activity can be simulated the model proposed and developed in Chap. 5 whose spectral characteristics of transient impulses have been gleaned from data. The proposed time series models are able to reproduce the transient behaviour of PD impulses that are randomly distorted by the impulse response of both the propagation channel and the receiver. Following the proposed model, a generalized PD impulse waveform is modelled numerically with an LTI filter, which is given by:

$$u_t = \sum_{i=1}^{p_u} \phi_i u_{t-i} + \varepsilon_t + \sum_{k=1}^{q_u} \psi_k \varepsilon_{t-k} \tag{7.3}$$

where u_t is the real-valued discrete-time impulse waveform from PD activity, u_{t-i} is the sample at discrete-time $t - i$. ϕ_i and ψ_k for any $i = \{1, \cdots , p_u\}$ and $k = \{1, \cdots , q_u\}$ are spectral characteristics of the impulse whose coefficients lie outside the unit circle. ε_t is the disturbance term which controls the impulsiveness and the duration of the distorted impulsive noise. This is modelled as a heteroskedastic Gaussian noise. In the presence of multiple PD sources, we assume that these parameters such as ψ_k and ϕ_i are *iid* random variables. These values are summarized in the set θ. The random parameter a and all of the random variables in θ are independent. Furthermore, ψ_k and ϕ_i are set such that the stability is guaranteed (see Chap. 5).

7.2.2.3   The Attenuation Factor

We assume that the antenna and PD sources are positioned to yield far-field conditions; in other words, PD sources must be positioned so that the distance source-receiver is, at least, greater than the wavelength of the antenna's receiver. As a result, the attenuation factor $c(\mathbf{r})$ in Eq. (7.1) is a decreasing function of the distance from the source and the receiver. For simplicity, $c(\mathbf{r})$ is approximated by the inverse of the source-receiver distance with an attenuation coefficient p, such that:

$$c(\mathbf{r}) = \frac{c_0}{r^p} \tag{7.4}$$

where $r = \|\mathbf{r}\|$ is the source-receiver distance and c_0 is a real positive random value induced by multipath effects. These parameters can be obtained by measuring the propagation channels in substations or any other environments where PD sources are observed.

7.2.3   Spatial and Temporal Distribution of PD Sources

The spatial and temporal distributions of PD sources are intrinsically linked to the number of pieces of HV equipment, their aging and the intensity of the electric field within PD sites. Let $\varXi$ be a finite space-time region in which the receiver observes such electromagnetic radiations emitted by PD sources inside the spatial region $\varUpsilon(R_1, R_2)$, where $R_1 \leq r \leq R_2$, and within a finite time interval $[0, T]$ where $\varXi = \varUpsilon(R_1, R_2) \times [0, T] \subset \varUpsilon \times [0, +\infty]$. In such instances, we shall approximate the noise process as follows: we assume that two PD sources cannot be located in the exact same position and that their emissions could not occur at the same exact time. In addition, it is also assumed that the number of PD events in the future is identically distributed over space and, independent of past events. From this approximation, we claim that the spatial and temporal distribution of PD sources follows a space-time Poisson point process [176–178]. Thus, the total number of impulses arriving at the receiver $N_I(t)$ inside the space-time region Ξ is a random variable following a space-time Poisson point process whose intensity is given by $\lambda(\mathbf{r}, t)$ where:

$$
\begin{aligned}
N_s &= \int_{\Xi} \lambda(t, \mathbf{r}) dt d\mathbf{r} \\
&= \int_0^T \int_{\Upsilon(R_1, R_2)} \lambda(t, \mathbf{r}) dt d\mathbf{r}
\end{aligned}
\tag{7.5}
$$

where N_s is the associated parameter of the Poisson distribution of PD sources inside Ξ, which is the average number of PD sources radiating transient impulses. The probability mass function of N_I is written as:

$$
Pr(N_I = k) = \frac{N_s^k}{k!} e^{-N_s}
\tag{7.6}
$$

for all $k = 0, 1, 2, 3, \cdots$. To yield mathematically tractable results, we assume the homogeneity of the space-time Poisson point process whose intensity $\lambda(t, \mathbf{r}) = \lambda$ is constant. Physically, the intensity λ is proportional to the number of pieces of HV equipment, their aging, the degradation of the electrical insulation systems, and the average intensity of the electric field within PD sites. Under such conditions, if the observation period is short, then the physical characteristics of HV installations does not vary over space and time.

From the homogeneity of the space-time Poisson point process, the average number of PD sources radiating transient impulses from Eq. (7.5) is written as:

$$
N_s = \lambda S
\tag{7.7}
$$

where S includes the volume of the spatial region and the time observation. We have restricted the spatial region to a finite sphere in the three-dimensional Euclidean space within a time observation $[0, T]$, $S = \frac{4}{3}\pi R^3 T$ where R is the radius of the finite sphere. Therefore, the parameter λ is the density of PD sources radiating impulses. Note that if the cyclostationary assumption must hold, then the density is a time-dependent function with respect to the periods of AC voltages. For simplicity, we assume that λ is constant.

The Poisson field of interferers allows us to formulate a generalized and physically complete noise model for substation environments, in which the physical characteristics of HV installations are grouped into a few statistical quantities. The proposed impulsive model is a filtered Poisson process whose typical response function $u(\theta, t, \mathbf{r})$ is a transient impulse induced by PD activity in substations. The spectral characteristics, amplitude, and durations can be captured from data (see Chap. 5). Moreover, the number of PD sources and their emissions can be controlled by the density parameter $\lambda(t, \mathbf{r})$ of the spatial and temporal Poisson point process.

7.3 Statistical Analysis

In this section, interesting statistical properties of the noise process can be derived based on practical assumptions. Eventually, they could be used to design a more robust receiver in impulsive noise environments, or in the development of new PD identification and detection methods for rapid diagnostics of electrical insulation systems in HV equipment via wireless IEDs. To do so, it is necessary to derive the first-order statistics of the resulting noise.

7.3.1 Probability Density Function

Taking advantage of the Poisson field of interferers, we derive the probability density function (PDF) of the noise of Eq. (7.2) via its characteristic function (c.f.). Inside the space-time region $\varXi$, the c.f. of the overall noise process X is given by:

$$\mathscr{M}_X(j\xi) = \sum_{k=0}^{\infty} \mathbb{E}\left[\exp(j\xi X)\right] P(k) \tag{7.8}$$

where $\mathbb{E}\left[\cdot\right]$ denotes the expectation and $P(k)$ is the probability of getting k PD impulses observed by the receiver in $\varXi$. Since we have assumed a homogeneous Poisson point process whose density parameter is constant λ, the average number of PD sources radiating transient impulses N_s is constant. Since X is filtered by a basic PD impulse waveform $u(\theta, t, \mathbf{r})$ with a real positive random amplitude a whose $\{a_i, \theta_i, t_i, \mathbf{r}_i\}$ is an *iid* random sequence for all $i = 0, 1, 2, \cdots$, and the positions of PD sources and their occurrences are uniformly distributed on any interval of space and time inside $\varXi$, the characteristic function is written as:

$$\mathscr{M}_X(j\xi) = \sum_{k=0}^{\infty} \left(\frac{1}{S}\int_{\varXi} \mathbb{E}\left[\exp(j\xi au(\theta, t, \mathbf{r}))\right] dt d\mathbf{r}\right)^k \frac{N_s^k}{k!} e^{-N_s} \tag{7.9}$$

where S includes the volume of the spatial region and the time observation. From the Taylor series expansion of an exponential function, Eq. (7.9) is written as:

$$
\begin{aligned}
\mathscr{M}_X(j\xi) &= \exp\left(\lambda \int_{\varXi} \left(\mathbb{E}\left[\exp(j\xi au(\theta, t, \mathbf{r}))\right] - 1\right) dt d\mathbf{r}\right) \\
&= \exp\left(\lambda \int_{\varXi} \left(\mathscr{Q}_{a,u}(j\xi) - 1\right) dt d\mathbf{r}\right)
\end{aligned} \tag{7.10}
$$

Note that $S = \int_{\varXi} dt d\mathbf{r}$. From Sect. 7.2, we have assumed that $\{a, \theta\}$ and $\{t, \mathbf{r}\}$ are independent and, therefore, all of these parameters are also independent. The PDFs of the real positive amplitude a and the set of random variables θ are denoted by

$p_a(a)$ and $p_\theta(\theta)$ respectively. Using the attenuation factor in Eq. (7.4) and since $\{a, \theta\}$ and $\{t, \mathbf{r}\}$ are independent, we write $\mathscr{Q}_{a,u}(j\xi)$ as follows [179, 180]:

$$\mathscr{Q}_{a,u}(j\xi) = \int_{\mathbb{R}^+} p_a(a) \int_\Theta p_\theta(\theta) \exp\left(j\xi \frac{c_0}{r^p} au(\theta, t)\right) d\theta\, da \qquad (7.11)$$

Assuming that c_0 is a positive random value with a PDF $p_{c_0}(c_0)$, for which the variable and all mentioned random variables are independent. Thus, $\mathscr{Q}_{a,u}(j\xi)$ can be written as:

$$\mathscr{Q}_{a,u}(j\xi) = \int_{\mathbb{R}^+} p_a(a) \int_\Theta p_\theta(\theta) \int_{\mathbb{R}^+} p_{c_0}(c_0) \exp\left(j\xi \frac{c_0}{r^p} au(\theta, t)\right) dc_0\, d\theta\, da$$
$$(7.12)$$

As an approximation, let the space-time region $\varXi$ go to infinity. Now, by taking the logarithm of the characteristic function of X in Eq. (7.10), and using the Fubini's theorem, we have [179, 180]:

$$\log \mathscr{M}_X(j\xi) = \lambda \int_{\mathbb{R}^+} p_a(a) \int_\Theta p_\theta(\theta) \int_{\mathbb{R}^+} p_{c_0}(c_0) \int_\varXi \left[\exp\left(j\xi \frac{c_0}{r^p} au(\theta, t)\right) - 1\right] d\mathbf{r}\, dt\, da\, d\theta\, dc_0$$

$$= \lambda \int_{\mathbb{R}^+} p_a(a) \int_\Theta p_\theta(\theta) \int_{\mathbb{R}^+} p_{c_0}(c_0) \int_0^\infty$$
$$\cdot \int_\Upsilon \left[\exp\left(j\xi \frac{c_0}{r^p} au(\theta, t)\right) - 1\right] d\mathbf{r}\, dt\, dc_0\, d\theta\, da$$
$$(7.13)$$

where $R_1 \to 0$, R_2 and $T \to \infty$. Since the spatial region is restricted to a sphere with a radius $R \to \infty$, Eq. (7.13) is can be written in spherical coordinates as:

$$\log \mathscr{M}_X(j\xi) = 4\pi\lambda \int_{\mathbb{R}^+} p_a(a) \int_\Theta p_\theta(\theta) \int_{\mathbb{R}^+} p_{c_0}(c_0) \int_0^\infty$$
$$\cdot \int_0^\infty \left[\exp\left(j\xi \frac{c_0}{r^p} au(\theta, t)\right) - 1\right] r^2 dr\, dt\, dc_0\, d\theta\, da$$
$$(7.14)$$

replacing g with $c_0 a r^{-p}$, the integration-by-substitution gives the following equation:

$$\log \mathscr{M}_X(j\xi) = \frac{4\pi\lambda}{p} \int_{\mathbb{R}^+} a^{3/p} p_a(a) \int_\Theta p_\theta(\theta) \int_{\mathbb{R}^+} c_0^{3/p} p_{c_0}(c_0) \int_0^\infty$$
$$\cdot \int_0^\infty \left[\exp\left(j\xi g u(\theta, t)\right) - 1\right] g^{-3/p-1} dg\, dt\, dc_0\, d\theta\, da$$
$$(7.15)$$

Hence, Eq. (7.15) is simplified as:

$$\log \mathscr{M}_X(j\xi) = \frac{4\pi\lambda}{p} \mathbb{E}\left[a^\alpha\right] \mathbb{E}\left[c_0^\alpha\right] \int_\Theta p_\theta(\theta) \int_0^\infty \int_0^\infty \mathscr{F}(\theta, t, g) dg\, dt\, d\theta$$
$$(7.16)$$

where $\alpha = 3/p$ is the stability index of the noise process and $\mathcal{F}(\theta, t, g)$ is given by:

$$\mathcal{F}(\theta, t, g) = \left[\exp\left(j\xi g u(\theta, t)\right) - 1\right] g^{-\alpha-1} \tag{7.17}$$

Note that Eq. (7.16) is valid if and only if $\mathbb{E}\left[a^{\alpha}\right]$ and $\mathbb{E}\left[c_0^{\alpha}\right]$ are finite and the integrand $\mathcal{F}(\theta, t, g)$ is absolutely integrable. Using Euler's formula, the integral over space is given by:

$$\int_0^{\infty} \mathcal{F}(\theta, t, g) dg = \int_0^{\infty} (\cos(bg) - 1) g^{-\alpha-1} dg + j \int_0^{\infty} \sin(bg) g^{-\alpha-1} dg$$

$$= \int_0^{\infty} -2\sin^2(bg/2) g^{-\alpha-1} dg + j \int_0^{\infty} \sin(bg) g^{-\alpha-1} dg \tag{7.18}$$

where $b = \xi u(\theta, t)$. In such instances, the integral on the left-hand side is absolutely integrable if $0 < \alpha < 2$, and the right-hand side is absolutely integrable if $0 < \alpha < 1$ [181]. In this case, the integral is absolutely integrable if the attenuation coefficient is $p > 3$ since $\alpha = 3/p$. If we assume that the noise process X has a symmetric probability distribution, then the characteristic function is real-valued. As a result, the integral is absolutely integrable if $p > 3/2$.

$$\int_0^{\infty} \mathcal{F}(\theta, t, g) dg = -\frac{\Gamma(1-\alpha)}{\alpha} |b|^{\alpha} \left(\cos\left(\frac{\pi}{2}\alpha\right) - j\,\mathrm{sign}(b)\sin\left(\frac{\pi}{2}\alpha\right)\right) \tag{7.19}$$

where $\Gamma(\cdot)$ is the gamma function, such that:

$$\Gamma(x) = \int_{\mathbb{R}^+} t^{x-1} e^{-t} dt \tag{7.20}$$

Finally, the logarithm of the characteristic function of X in Eq. (7.16) is given by:

$$\log \mathcal{M}_X(j\xi) = -\sigma |\xi|^{\alpha} \left(1 - j\beta\,\mathrm{sign}(\xi)\tan(\pi\alpha/2)\right) \tag{7.21}$$

where

$$\sigma = \frac{4\pi\lambda\Gamma(1-\alpha)\cos(\pi\alpha/2)}{p\alpha} \mathbb{E}\left[a^{\alpha}\right] \mathbb{E}\left[c_0^{\alpha}\right] \int_{\Theta} p_{\theta}(\theta) \int_0^{\infty} |u(\theta, t)|^{\alpha} dt\, d\theta \tag{7.22a}$$

$$\beta = \mathrm{sign}(u(\theta, t)) \tag{7.22b}$$

where $\sigma > 0$ and $\beta \in [-1, 1]$. The characteristic function of X has the form of the c.f. of the α-stable, if $\mathbb{E}\left[a^{\alpha}\right]$, $\mathbb{E}\left[c_0^{\alpha}\right]$ and $\int_0^{\infty} |u(\theta, t)|^{\alpha} dt$ are finite. Therefore, the PDF of X can be approximated as an α-stable distribution. Note that if X has a symmetric distribution, then the logarithm of the c.f. in Eq. (7.21) is simply:

$$\log \mathcal{M}_X(j\xi) = -\sigma |\xi|^{\alpha} \tag{7.23}$$

By using the Poisson field of interferers in which PD sources radiate transient impulsive waveforms in a realistic scenario, we have demonstrated that the PDF of X can be approximated as an α-stable distribution, under certain conditions. This is characterized by a heavy-tailed behaviour induced by the stability index $0 < \alpha < 2$ and a possible skewness (i.e. the probability distribution is asymmetric) induced by β. However, in practice, the noise process consists of impulsive noise and an additive background noise, including thermal noise, in the receiver. Hence, the resulting noise process is given by:

$$X = \sum_{k=1}^{N_I} a_k u(\theta_k, t_k, \mathbf{r}_k) + n(t) \tag{7.24}$$

where $n(t)$ is background noise, which is a spatially isotropic noise process. Assuming that the central limit theorem holds for the background noise (i.e. non-impulsive noise), we consider this to be a Gaussian noise. Under this condition, the logarithm of the c.f. of the overall noise process is written as:

$$\log \mathscr{M}_X(j\xi) = -\sigma |\xi|^\alpha \left(1 - j\beta \operatorname{sign}(\xi) \tan(\pi\alpha/2)\right) - \sigma_n^2 \xi^2 \tag{7.25}$$

where σ_n^2 is the variance of the background noise. Note that the impulsive noise process and the background noise are independent. As a result, the PDF of X is given by:

$$f_X(x) = \frac{1}{2\pi} \int_{\mathbb{R}} \mathscr{M}_X(j\xi) e^{-j\xi x} d\xi \tag{7.26}$$

The $f_X(x)$ of the noise process might be difficult to derive in closed-form because α-stable PDFs cannot be written analytically except when $\alpha = 2, 1$, and $1/2$, which are Gaussian, Cauchy and Levy distributions respectively. When the background noise is high, the PDF might be approximated by a Gaussian distribution when $\sigma_n^2 \gg \sigma$. Assuming that the PDF of the noise process is α-stable, the statistical analysis of the amplitude probability distribution, as well as its tails and moments, is well-established. Extensive research covering the characterization and implementation of α-stable noise processes has been provided in the literature [77, 93, 138, 139, 182]. This analysis is summarized in the following paragraphs. We will show that the high-order moments of the α-stable distributions are not finite. In particular, the second-order moment is infinite. Unless fractional lower-order moments are used, it is inappropriate to use an infinite variance in signal processing problems [183]. Fortunately, since the noise process is based on the Poisson field of interferers, Campbell's theorem allows for the derivation of high-order moments of the noise process, which are finite under certain conditions.

7.3.2 Probability Distribution

The probability distribution is the cumulative distribution function (CDF) of the noise. This is characterized by the PDF of the noise, such that:

$$F_X(x) = P(X \leq x) = \int_{-\infty}^{x} f_X(\xi)d\xi \tag{7.27}$$

However, since we have demonstrated that the PDF is an α-stable distribution, a closed-form is not available when $0 < \alpha < 2$ and particularly when $\alpha \neq 0.5$, 1 and 2. Nevertheless, the CDF is determined by α and β namely; if $\beta = 0$, then the CDF is symmetric around $x = 0$. When $\beta > 0$, the CDF is right-skewed, i.e. the right-tail of the distribution is heavier than the left-tail and $P(X > x) > P(X < -x)$. On the other hand, the CDF is left-skewed, which means that the left-tail of the distribution is heavier than the right-tail and $P(X > x) < P(X < -x)$ when $\beta < 0$. If α decreases, then the tail probabilities increase.

7.3.3 Tails and Moments

The tail distribution is the complementary cumulative distribution function (CCDF) of the noise process. This is defined as:

$$\bar{F}_X(x) = P(X > x) = 1 - F_X(x) \tag{7.28}$$

In [139], authors have shown that tail behaviour is determined by α and β. When $0 < \alpha < 2$ and $-1 < \beta \leq 1$, then as $x \to \infty$ the tail approximation is given by:

$$P(X > x) \sim \sigma^\alpha \sin(\pi\alpha/2) \frac{\Gamma(\alpha)}{\pi}(1 + \beta)x^{-\alpha} \tag{7.29}$$

However, when $0 < \alpha < 2$ and $-1 \leq \beta < 1$, then as $x \to \infty$ the lower tail approximation is given by:

$$P(X < -x) \sim \sigma^\alpha \sin(\pi\alpha/2) \frac{\Gamma(\alpha)}{\pi}(1 - \beta)x^{-\alpha} \tag{7.30}$$

For any value of $\alpha < 2$ and $-1 < \beta < 1$, the PDF and the CCDF are asymptotically power laws. However, when $\beta = -1$ or 1, the tails of the distributions are not asymptotically power laws.

7.3.3.1 Moments of α-Stable Distributions

The moments of the probability distribution is given by:

$$\mu_k = \mathbb{E}\left[X^k\right]$$
$$= \int_{\mathbb{R}} x^k f_X(x) dx \qquad (7.31)$$

where μ_m is the kth moment of the probability distribution. Note that if the moments are finite $\forall k < \infty$, then the mean, the variance and the kurtosis can be measured. Unfortunately, [139, 183] have indicated that all moments do not exist (i.e. μ_k is not finite) when the distribution is α-stable. In particular if $0 < \alpha < 2$, then:

$$\mathbb{E}\left[X^k\right] = \begin{cases} \infty, & \text{if } k \geq \alpha. \\ \mu_k < \infty, & \text{if } 0 < k < \alpha. \end{cases} \qquad (7.32)$$

In other words, for $0 < \alpha \leq 1$, first- or higher-order moments are not finite; for $1 < \alpha < 2$, the first and all of the kth moments are finite when $k < \alpha$. In particular, α-stable distributions have infinite variance (i.e. $\mu_2 - \mu_1^2 = \infty$). Under such conditions, higher-order moments such as the second-order and/or the fourth-order moment cannot be exploited in practice. A study reported in [183] has derived fractional lower-order moments (i.e. $\mu_k < \infty$ when $0 < k < \alpha$) for practical engineering applications. Unfortunately, the authors have stated that these lower-order moments are much harder to work with than second- and higher-order moments because they introduce non-linearity to even linear problems.

7.3.3.2 Moments of Shot-Noise Processes

Fortunately, since the proposed model is based on the Poisson field of interferers and the noise process X is a shot-noise process filtered by a basic transient impulsive waveform, Campbell's theorem allows for the derivation of first- or higher-order moments [179]. Assuming that the impulsive noise process and the background noise are independent, statistical moments are written as:

$$\mathbb{E}\left[X^k\right] = \lambda\mathbb{E}\left[(au(\theta, t, \mathbf{r}))^k\right] + \mathbb{E}\left[n^k(t)\right] \qquad (7.33)$$

Using the proposed basic waveform $au(\theta, t, \mathbf{r})$ and assuming that $\{a_i, \theta_i, t_i, \mathbf{r}_i\}$ is an *iid* random sequence for all $i = 0, 1, 2, \cdots$, we write the statistical moments as:

$$\mathbb{E}\left[X^k\right] = \lambda\mathbb{E}\left[a^k\right]\mathbb{E}\left[u^k(\theta, t, \mathbf{r})\right] + \mathbb{E}\left[n^k(t)\right]$$
$$= \lambda\mathbb{E}\left[a^k\right]\mathbb{E}\left[c_0^k\right]\int_0^T \mathbb{E}\left[u^k(\theta, t)\right]\int_{\Upsilon(R_1, R_2)} r^{-pk} d\mathbf{r}\, dt + \mathbb{E}\left[n^k(t)\right]$$
$$\qquad (7.34)$$

The noise process X has finite first- or higher-order moments if and only if $\mathbb{E}\left[a^k\right]$, $\mathbb{E}\left[c_0^k\right]$ and $\mathbb{E}\left[n^k(t)\right]$ have finite moments and the integrands $\mathbb{E}\left[u^k(\theta, t)\right]$ and r^{-pk} are integrable over time and space respectively. In particular, by examining the integral over space in spherical coordinates, we have:

$$\int_{\Upsilon} r^{-pk} d\mathbf{r} = 4\pi \int_{R_1}^{R_2} r^{-(p-2)k} dr \tag{7.35}$$

Therefore, the noise process X has no finite first- or higher-order moments if and only if $R_1 = 0$ and $p - 2 \geq 1/k$ or if $R_2 = \infty$ and $p - 2 \leq 1/k$ [184]. In practice, the receiver does not detect PD sources when $R_2 = \infty$ and since we have assumed that the receiver is located at the origin of the sphere, there are no PD sources located at this position. Note that R_1 needs to be much greater than the wavelength of the antenna, thereby establishing far-field conditions, which renders the attenuation factor in Eq. (7.4) valid. If far-field conditions hold, then the integrand r^{-pk} is integrable over space $\forall k < \infty$. Moreover, since the waveform $u(\theta, t)$ observed by the receiver is a transient impulse with a finite energy and duration, then the integrand $\mathbb{E}\left[u^k(\theta, t)\right]$ is integrable over time $\forall k < \infty$.

7.3.4 A Summary of Important Findings

We have proposed a generalized, complete, and physically-coherent radio-noise model for substation environments with respect to the induced EMIs created by the presence of multiple PD sources in HV equipment. Based on practical assumptions, we have assumed that PD sources are distributed according to a space-time Poisson point process (the Poisson field of interferers). Accordingly, the basic transient impulsive waveform, $au(\theta, t, \mathbf{r})$, which is generated by these sources, has been specified according to the far-field wave propagation and physical characteristics of impulses such as amplitude, durations, and spectra.

This proposed model is a filtered Poisson point process, in which the number of PD sources per unit volume or unit area, as well as the number of emissions per unit time per source, can be set by a single density parameter λ. Although the cyclostationary process has been neglected in this work, one could add a time-dependent function into the parameter λ to conduct such processes. The relative intensity of the impulses can be derived using the far-field wave propagation and the intensity of radiations.

Taking advantage of the Poisson field of interferers, the PDF of the resulting noise X has been derived from the characteristic function $\mathscr{M}_X(j\xi)$. Under practical and realistic assumptions, we have demonstrated that the PDF can be approximated by an α-stable distribution. However, all moments of these distributions do not exist. In particular, the noise power level (second-order moment) and the degree of impulsiveness (fourth-order moment) are not finite. As a result, signal processing methods based on these statistical moments cannot be used for detection, estimation or identification. Fortunately, the Campbell's theorem allows for the derivation of first- or higher-order moments in closed-form. These moments are finite if and only if first- and higher-order moments of the basic transient impulsive waveform and its random parameters are finite. In particular, by examining the integral obtained in the spatial domain, we have found that first- and higher-order moments do not exist when PD sources are located closed to the antenna position. Theoretically, such conditions do not hold, because the attenuation factor used in Eq. (7.4) is valid, if and only if the far-field conditions are established.

The noise generation of the proposed model is physically coherent compared to other statistical models using first-order statistics. This is because the spectra, occurrences and durations of the impulsive component, as well as background noise, are taken into account. Moreover, the first-order statistics, such as probability distributions and moments, can be linked to the physical process of the noise, which can, in turn, be utilized for performing communications analyses and designing and optimizing communication systems in substations.

7.4 Experimental and Simulation Results

In this section, a simple procedure for estimation is presented in order to reproduce such noise processes in substations. Then, the estimated parameters will be used, by which the effectiveness of the proposed approach will be shown by comparing measurement and simulation results in terms of first- and second-order statistics. In addition, since α-stable and Middleton Class A impulsive noise models are commonly utilized in substations environments, they will be used for comparison [45, 71, 72, 102].

7.4.1 Measurements in Substations

Using the measurement setup that has been employed in Chaps. 3–5, the antenna is surrounded by HV equipment, by which multiple PD sources can be detected. By setting a time observation of $T = 5\,\mu s$, many impulses from PD activity are captured via an oscilloscope. 400 temporal waveforms have been recorded in the frequency range of 800 MHz to 5 GHz. PD sources are characterized by any RF gain in the measurement setup by removing the antenna factor. We have used a frequency mixer with a local oscillator of $f_0 = 800$ MHz in order to yield a baseband representation of RF signals. The EM radiations are measured in terms of electric strength (V/m).

Since the antenna is positioned in a specific position within a substation and we do not have any information on the location of PD sources, the noise process X is written as a temporal function, such that:

$$X(t) = \sum_{k=1}^{N_I} a_k u_k(\theta_k, t) + n(t) \tag{7.36}$$

where N_I is the number of observed impulses and a_k is a random amplitude including the attenuation induced by wave propagation and the amplitude generated by a PD source. $n(t)$ is the background noise.

7.4.2 A Procedure for Estimation

Based on the proposed characterization process in Chap. 3, a simple estimation procedure of the noise process can be proposed, in which the estimated parameters can be utilized in the proposed generalized impulsive noise model. The estimation procedure can be presented as follows:

- assuming $n(t)$ is a Gaussian noise, record waveforms when PD impulses have not occurred. Then, estimate the level of the background noise, such that:

$$\hat{\sigma}_n^2 = \text{Var}\,[n(t)] \tag{7.37}$$

where $\text{Var}\,[\cdot]$ is the estimated variance;
- in the presence of PD impulses, count the number of impulses that have occurred. Hence, the parameter density λ can be estimated by the average number of impulses occurring over time observation:

$$\hat{\lambda} = \frac{\bar{N}_I}{T} \tag{7.38}$$

where $\hat{\lambda}$ is the estimated density parameter and $\bar{N}_s$ is the average number of impulses.
- detect each impulse and estimate the spectral characteristics via the proposed time series models with heteroskedasticity. Note that the duration of an impulse can be estimated using ARCH, or EGARCH models. Next, check the goodness-of-fit of these models via the estimation procedure detailed in Chap. 5;
- the amplitude a can be provided by measuring the variance of each PD impulse, such that:

$$\hat{a}_k^2 = \text{Var}\,[a_k u(\theta_k, t_k)]$$
$$= \text{Var}\,[a_k] \int_0^{T_u} |u^2(\theta_k, t_k)| dt \tag{7.39}$$

where $\hat{a}_k^2$ is the estimated square of the amplitude and T_u is the duration of an impulse. Therefore, $\hat{a}_k$ is obtained via the square-root of $\hat{a}_k^2$. Next, plot the histogram of $\hat{a}_k$ or $\hat{a}_k^2$ and estimate the probability density.

In the measurement campaign, the estimated background noise level is $\hat{\sigma}_n^2 = -65.35$ dBW/m^2, the average number of PD emissions is $\bar{N}_s = 2.46$ (i.e. the density is $\hat{\lambda} = 4.92 \times 10^{-5}$ s^{-1}), and the power density of PD impulses $\hat{a}_k^2$ follows an exponential distribution whose average power density is $\bar{a}_k^2 = -51.96$ dBW/m^2.

7.4.3 Measurement-Simulation Comparison

Measurement and simulation results are presented for comparison. The goodness-of-fit is measured to assess the effectiveness of the approach.

7.4.3.1 First-Order Statistics

Using the proposed generalized model, the estimated parameters have been used in order to reproduce the impulsive noise process in substations. The adequacy of the model is discussed via fist-order statistics from simulation and measurement results. α-stable and Middleton Class A distributions are also provided. The empirical probability distributions (PDFs, CDFs, and CCDFs) are presented to assess the goodness-of-fit of these models.

In Fig. 7.1, a typical waveform is obtained by simulation using a second-order of autoregressive time series model with heteroskedasticity. The LTI filter reproduces the spectrum of an impulse. Distortions and the duration are generated by the disturbance term. Using the estimated parameters, the noise process X via the proposed impulsive noise model is depicted in Fig. 7.2.

As a comparison, 400 waveforms have been simulated to yield the overall empirical PDFs, CDFs, and CCDFs. In addition, parameters of α-stable and Middleton Class A have been estimated using statistical methods as developed by [73, 75, 91, 185]. Then, probability distributions are plotted and the goodness-of-fit of the impulsive noise model are measured using the Kullback-Leibler (KL) divergence, and the Kolmogorov-Smirnov (KS) test. The empirical PDFs and CCDFs of measurements and impulsive noise models are depicted in Fig. 7.3. In the presence of impulsive noise, the PDF's behaviours are heavy-tailed, and the probability of having a strong amplitude ($X > |0.05|$ V/m) represents to the impulsive component. The probability densities are symmetric (i.e. $f_X(x) = f_X(-x)$), and the tail can be described approximately as a power-law, such that $\bar{F}_X(x) \sim |x|^{-\nu}$ as $x > 0.05$ V/m, where ν is an exponent characterizing the decay. An accurate impulsive noise model can be defined by its ability to reproduce the decay in the probability distribution. By comparing PDFs, however, these impulsive noise models can reproduce the heavy-tailed behaviour; our generalized impulsive

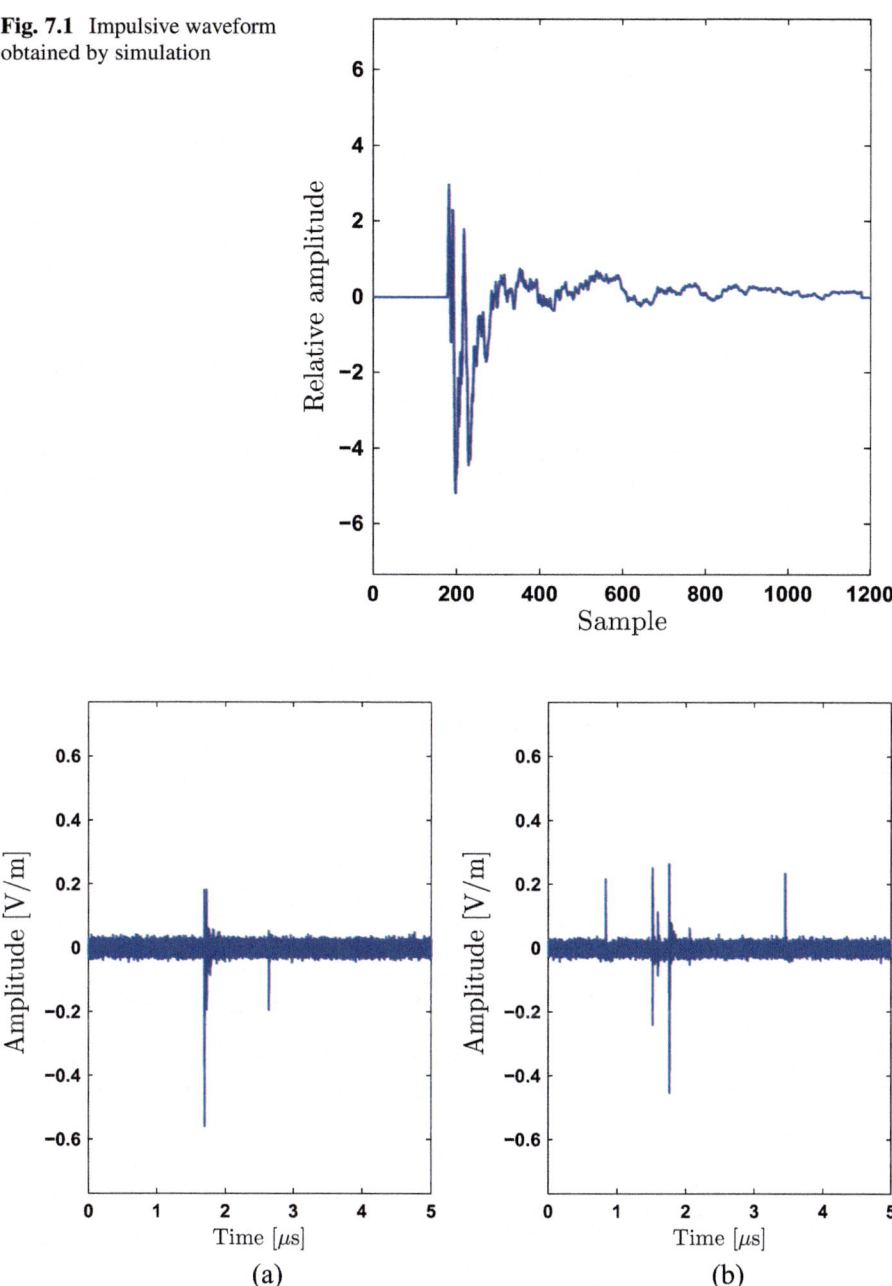

Fig. 7.1 Impulsive waveform obtained by simulation

Fig. 7.2 Physical shot-noise process X. (**a**) Measurement. (**b**) Simulation

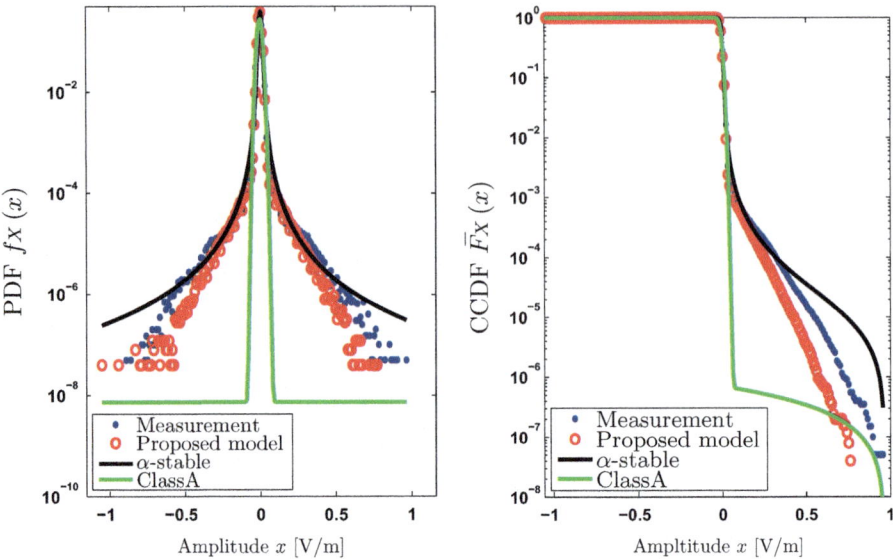

Fig. 7.3 Probability distributions of the noise process

Table 7.1 The goodness-of-fit: measurement vs. impulsive noise models

Test statistics	Proposed model	α-Stable	Middleton class A
D_{KL}	0.0140	0.0437	0.0756
D_{KS}	0.0306	0.0547	0.1294

noise model is therefore more accurate compared to the α-stable and the Middleton Class A distribution. Indeed, the proposed model has the smallest value in terms of KL-divergence (see Table 7.1). The KS-test value in the proposed model has again the smallest value; the p-values are 0.622, 0.5181, and 2.0×10^{-4} for the proposed model, α-stable, and Middleton Class A respectively. As a result, the KS-test rejects the null hypothesis that measurement results and Middleton Class A noise model come from the same distribution at a 95% confidence interval. The test does not reject the null hypothesis for the other impulsive noise models. In addition, by comparing CCDFs, the tail of the proposed model is closer than other impulsive noise models.

7.4.3.2   Second-Order Statistics

Although α-stable and Middleton Class A noise models might be adequate for noise processes in substations, computer methods for reproducing noise samples are limited because the resulting random process X is independent and identically distributed so that the impulsive component is modelled as a single-noise sample.

As the noise process X is a collection of independent random variables, the resulting autocorrelation function $R_X(\tau)$ is defined as:

$$R_X(\tau) = \mathbb{E}\left[X^2\right]\delta(\tau) \tag{7.40}$$

where $\delta(\tau)$ is the Dirac impulse function. Under such a condition, the power spectral density (PSD) of such noise processes is constant over all frequencies. In particular, an α-stable noise process does not have a PSD because $R_X(0) = \mathbb{E}\left[X^2\right]$, and we have shown that its second-order moment does not exist. In practice, impulsive noise components are characterized by transient effects whose spectra have a form of $\sim(f - f_0)^{-\gamma}$. From Eq. (7.36), Campbell's theorem allows us to define the autocorrelation function and the power spectral density of the noise process X, such that:

$$
\begin{aligned}
R_X(\tau) &= \lambda R_u(\tau) + R_n(\tau) \\
&= \lambda\mathbb{E}\left[a^2\right]\mathbb{E}\left[u(\theta, t)u(\theta, t + \tau)\right] + \mathbb{E}\left[n(t)n(t + \tau)\right]
\end{aligned}
\tag{7.41}
$$

where τ is a temporal lag. If $R_u(\tau)$ exists for all value of τ, then the PSD of X is defined by the Fourier transform of $R_X(\tau)$ [186]:

$$
\begin{aligned}
S_{xx}(f) &= \int_{\mathbb{R}} R_X(\tau)e^{-j2\pi f\tau}d\tau \\
&= \lambda\mathbb{E}\left[a^2\right]\int_{\mathbb{R}}(R_u(\tau) + R_n(\tau))e^{-j2\pi f\tau}d\tau \\
&= \lambda\mathbb{E}\left[a^2\right]S_{uu}(f) + S_{nn}(f)
\end{aligned}
\tag{7.42}
$$

where $S_{uu}(f)$ and $S_{nn}(f)$ are PSDs of the transient PD impulse and background noise respectively. Since $S_{uu}(f)$ has a form of $\sim(f - f_0)^{-\gamma}$, the PSD $S_{xx}(f)$ has a form of:

$$S_{xx}(f) \sim (f - f_0)^{-\gamma} + S_{nn}(f) \tag{7.43}$$

The behaviour of such PSDs can be emulated via LTI filters. By using the periodogram, the average PSD of the noise process has been estimated, as depicted in Fig. 7.4. The measured noise process contains transient PD impulses and background noise. Since we have used a frequency mixer with a local oscillator at f_0, the average PSD has a form of $\sim f^{-\gamma}$. On average, the background noise consists of thermal noise, RF communications and harmonics caused by interleaving artefacts and clock feedthrough from the oscilloscope. The average PSD of the emulated noise process matches experimentation. A more accurate noise model can be achieved by modeling RF communications and harmonics from the oscilloscope. This case is beyond the scope of this work. The PSD of impulses has been extracted from the ambient noise using a denoising process, as depicted in Fig. 7.4b. Again, by comparing the average PSDs, simulation results match experimentation.

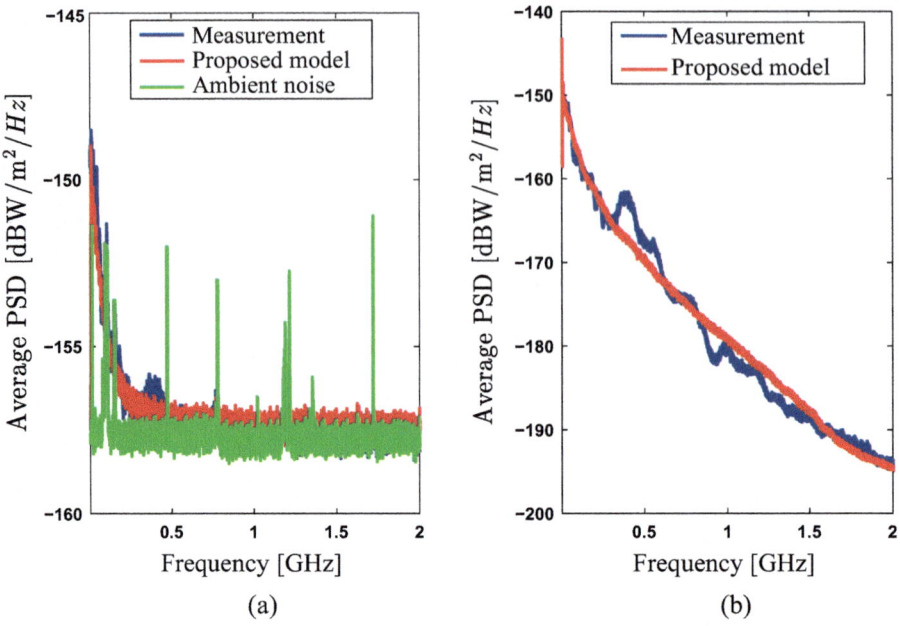

Fig. 7.4 Second-order statistics: average PSD. (**a**) Raw PSD. (**b**) Denoised PSD

Using filtered Poisson processes may be appropriate when impulsive noise has a spectrum of $\sim f^{-\gamma}$ rather than using memoryless impulsive noise models that generate *iid* random noise. Based on the proposed estimation methods, the effectiveness of the approach has been shown for substation environments when impulses are generated by PD sources. Practical applications can be implemented based on advanced signal processing methods for detecting, estimating, and identifying any desired RF signals from RF communications and/or PD sources.

7.5 A Rapid Identification of PD Sources Using Blind Source Separation

In this section, we proposed a practical application for a rapid identification of PD sources by using blind-source separation and calculating the fourth-order moment. Spatial second- and higher order statistics have been exploited through multiple antennas (array of antennas) combined by a measure of the degree of impulsiveness at each antenna to estimate the number of PD sources in HV equipment.

7.5.1   Motivation

Recent advances in wireless sensor networks in substations can provide significant improvements for protection, control, automation and monitoring. One example of remote monitoring applications is a rapid insulation diagnosis in power equipment using wireless IEDs. PD activities can cause irreversible damage and possible failure of electrical insulation systems. Insulation performances and lifespan can be evaluated by measuring the number of PD sources in power equipment. Electromagnetic radiations can be detected by wireless devices for remote monitoring applications. The resulting signal is highly impulsive when the spectrum can cover very large frequency bands (above few GHz) [124, 141].

A technique for a rapid identification of PD sources is proposed. This can be implemented in a low cost wireless IED for remote monitoring applications. By using multiple antennas, PD sources can be separated based on BSS techniques via generalized eigenvalue decomposition presented in [187]. Furthermore, the presence of significant impulsive events produces heavy-tailed distribution when the excess kurtosis is greater than zero. Therefore, the number of PD sources can be estimated by measuring the excess kurtosis of the demixed signals at each antenna. To demonstrate the efficiency and the performance of the proposed method, we use a generic and realistic impulsive noise model, with which impulsive waveforms are generated by autoregressive (AR) models. Their parameters can be estimated from various measurement campaigns in substations. Discharge sources are modelled as a spatial PPP. One can control physical parameters such as the density of PD sources and the average intensity of impulsive component over background noise.

7.5.2   System Model

We consider N unknown PD sources written as a $N \times 1$ vector $\mathbf{u}$ of a collection of impulsive waveforms as:

$$\mathbf{u} = [u_1(\theta, t)u_2(\theta, t) \cdots u_N(\theta, t)]^T \tag{7.44}$$

where $u_i(t)$ are complex-valued signals emitted by the ith PD sources. The receiver has M temporal samples $\mathbf{x}$ into a $M \times N$ complex-valued mixing channel $\mathbf{H}$, such that:

$$\mathbf{x} = \mathbf{H}\mathbf{u} \tag{7.45}$$

$$\mathbf{y} = \mathbf{x} + \mathbf{n} \tag{7.46}$$

where $\mathbf{n}$ is an additive background noise generated by thermal noise and ambient noise in a substation. Note that no precise knowledge is available regarding either the mixing channel or the sources. To recover discharge sources from observation, we need to find $\mathbf{G}$, an inverse matrix, such that:

$$\hat{\mathbf{u}} = \mathbf{G}^*\mathbf{y} \tag{7.47}$$

where $\mathbf{G}^*\mathbf{H} = \mathbf{I}$ and $\mathbf{G}^*$ is the conjugate transpose of $\mathbf{G}$. The number of PD sources can be estimated from the demixed signals by measuring the excess kurtosis given by the fourth moment about the mean minus three:

$$\kappa_i = \frac{\mathbb{E}\left[(\hat{u}_i(t) - \mu_i)^4\right]}{(\mathbb{E}\left[(\hat{u}_i(t) - \mu_i)^2\right])^2} - 3 \tag{7.48}$$

where μ_i is the mean of the ith demixed signals. In the presence of significant impulsive signals, the distribution is heavy-tailed where $\kappa_i > 0$. Thus, the estimated number of PD sources $\hat{N}$ is given by summing the number of dimensions i, for which κ_i is greater than a given threshold $T_h > 0$ to be defined later:

$$\hat{N} = \sum_i (\kappa_i > T_h) \tag{7.49}$$

An overview of the system model is depicted in Fig. 7.5, where the receiver estimates the number of PD sources in power equipment for remote monitoring applications.

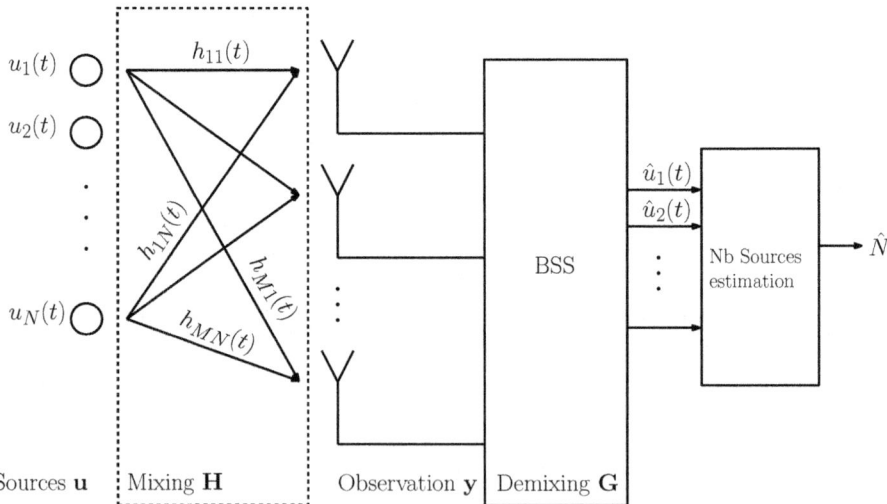

Fig. 7.5 Overview of the system model

7.5.3   Blind Source Separation via Generalized Eigenvalue Decomposition

Partial discharge phenomena are stochastic processes where the pulse height, event and spectrum depend on various physical parameters such as electric field intensity, free electron rate, and ageing mechanism [55]. Hence, in the presence of multiple PD sources, it is reasonable to consider that sources are independent, or at least decorrelated. Moreover, according to the physical characteristics of radiated radio frequency (RF) signals from PD activity measured by [124, 141], one can assume that PD are non-Gaussian. During measurement campaigns in substations, these impulsive signals have transient behaviour with damped oscillation. Therefore, sources may be non-white and/or non-stationary processes.

According to [187], two conditions are sufficient for source separation via generalized eigenvalue decomposition: (a) if sources are independent or decorrelated, the covariance matrix $\mathbf{R_y}$ can be written as:

$$\begin{aligned} \mathbf{R_y} &= \mathbb{E}\left[\mathbf{yy}^*\right] \\ &= \mathbf{R_x} + \mathbf{R_n} \\ &= \mathbf{HR_uH}^* + \mathbf{R_n} \end{aligned} \tag{7.50}$$

where $\mathbf{R_u}$ and $\mathbf{R_n}$ are diagonal. Therefore, $\mathbf{R_x}$ is also diagonal. Assuming the background noise is modelled as a circular complex Gaussian noise, $\mathbf{R_n} = \sigma^2\mathbf{I}$. (b) If PD sources are non-Gaussian, non-stationary, or non-white, there exists $\mathbf{Q_u}$ which has the same diagonalization property such that:

$$\mathbf{Q_x} = \mathbf{HQ_uH}^* \tag{7.51}$$

From Eqs. (7.46) and (7.51), we write:

$$\mathbf{Q_y} = \mathbf{Q_x} + \mathbf{Q_n} \tag{7.52}$$

where $\mathbf{Q_n}$ is also diagonalizable. From these two conditions, generalized eigenvalue decomposition can be used for source separation [187]. Indeed, in Eqs. (7.50) and (7.52), by multiplying them by $\mathbf{G}$ and Eq. (7.52) by $\Lambda = \mathbf{R_xQ_x^{-1}} + \mathbf{R_nQ_n^{-1}}$, we have:

$$\mathbf{R_yG} = \mathbf{Q_yG}\Lambda \tag{7.53}$$

From statistical assumptions, $\mathbf{Q}$ can have various forms. By assuming that sources are:

• non-stationary and decorrelated [188]; we have:

$$\mathbf{Q_{y,1}} = \mathbf{R_y} = \mathbb{E}\left[\mathbf{yy}^*\right] \tag{7.54}$$

- non-white and decorrelated [189]; we have:

$$\mathbf{Q}_{\mathbf{y},2} = \mathbf{R}_{\mathbf{y}}(\tau) = \mathbb{E}\left[\mathbf{y}(t)\mathbf{y}^*(t+\tau)\right] \tag{7.55}$$

where τ is a time delay. In this work, we consider that $\tau = 1$ sample.
- non-Gaussian and independent [190]; we have :

$$\mathbf{Q}_{\mathbf{y},3} = \mathbb{E}\left[\mathbf{y}^*\mathbf{y}\mathbf{y}\mathbf{y}^*\right] - \mathbf{R}_{\mathbf{y}}\mathrm{Tr}(\mathbf{R}_{\mathbf{y}}) - \mathbb{E}\left[\mathbf{y}\mathbf{y}^T\right]\mathbb{E}\left[\bar{\mathbf{y}}\mathbf{y}^*\right] - \mathbf{R}_{\mathbf{y}}\mathbf{R}_{\mathbf{y}} \tag{7.56}$$

where $\bar{\mathbf{y}}$ is the conjugate of $\mathbf{y}$ and $\mathrm{Tr}(\mathbf{R}_{\mathbf{y}})$ is the trace of $\mathbf{R}_{\mathbf{y}}$.

The inverse matrix $\mathbf{G}$ is given by the generalized eigenvalue of the matrices $\mathbf{R}_{\mathbf{y}}$ and $\mathbf{Q}_{\mathbf{y}}$.

7.5.4 Simulation and Results

Performance of the proposed technique is provided using a Monte Carlo simulation, in which PD sources are simulated based on spatial PPP and the number of sources is a random variable unknown to the receiver. The average number of sources per unit volume or surface is given by λ_s. Over a long observation time, sources emit impulsive discharges when their events follow a temporal PPP. The average number of emissions per unit time per sources is given by λ_e.

Transient impulsive signals from discharge activity are simulated by using autoregressive (AR) models, such that a single discrete-time impulse u_t is written as:

$$u_t = \sum_{i=1}^{p_u} \phi_i u_{t-i} + \varepsilon_t \tag{7.57}$$

where coefficients $\{\phi_i\}$ are parameters of the model which can be obtained from measurements in various substations. ε_t is a Gaussian noise process in which the variance is non-constant over time in order to take into account the impulse duration [141]. The pulse height is an exponential random variable for which the average intensity of impulsive component over background noise is given by $\Gamma = \sigma_u^2/\sigma_n^2$.

In this work, the receiver operates at an average rate of $\lambda_s = 4$ PD sources per unit volume or surface. For an observation time of 5×10^4 samples, the average number of emissions is $\lambda_e = 5$ discharges per unit time per source. We use the second order of the AR model where $p_u = 2$ in Eq. (7.57). Parameters are obtained from measurements in various substations.

For BSS, we use these three forms of $\mathbf{Q}_{\mathbf{y}}$ to recover PD sources. Their potential to estimate the correct number of sources is compared via their probability of error P_e (the estimated number of PD sources is not correct). To limit the probability of a false alarm, the estimated number of sources $\hat{N}$ is determined when the

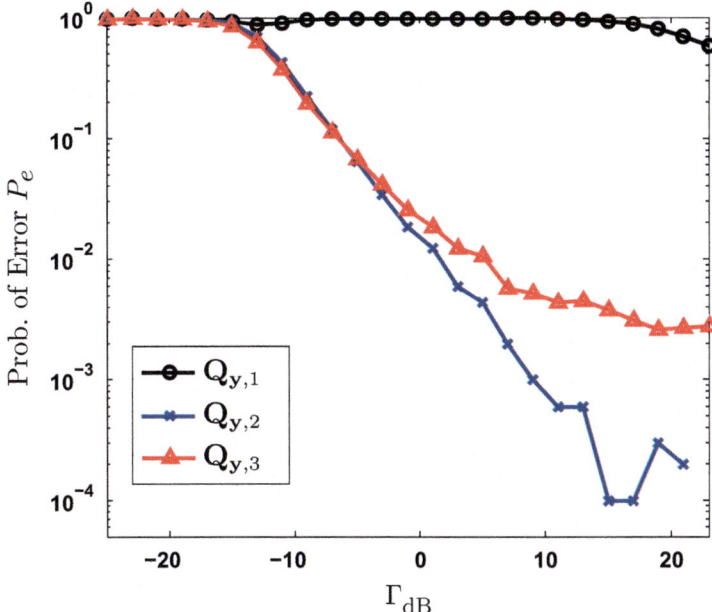

Fig. 7.6 Probability of error vs. Γ_{dB}. Based on PPP model: Avg. PD source density $\lambda_s = 4$ sources per unit volume or surface, avg. emission density $\lambda_e = 5$ discharges per unit time per sources, NB. obs. $M = 20 > N$

threshold $T_h = 1$ according to Eq. (7.49). Under this condition, the excess kurtosis is zero because the background noise is modelled as a Gaussian noise. Hence, the probability of a false alarm is $P_{fa} = 0$. The mixing channel **H** is a $M \times N$ circular complex Gaussian noise. The probability of error P_e, is obtained from a Monte Carlo simulation of 15,000 simulations.

The performance of the receiver is plotted in Fig. 7.6 where the probability of error P_e is given for different values of Γ and different forms of $\mathbf{Q_y}$. The number of PD sources is a random variable which is not known and not available to the receiver. The number of observations is $M = 20$, and thus always greater than the number of sources ($M > N$). By applying $\mathbf{Q_{y,2}}$ and $\mathbf{Q_{y,3}}$ for BSS, the probability of error P_e decreases when the average intensity of the impulsive component is higher than the background noise. However, P_e is high and nearly constant when $\mathbf{Q_{y,1}}$ is applied. As a result, the number of PD sources can be determined with low probability of error if we assume that the sources are non-white and decorrelated, or non-Gaussian and independent. For high values of Γ, where discharges are more significant, the receiver can estimate the exact number of PD sources when $\mathbf{Q_{y,2}}$ is applied.

Figure 7.7 shows the performance of the receiver for a given number of observations where only $\mathbf{Q_{y,2}}$ is applied. The probability of error P_e is high when the number of observations is lower than the number of PD sources ($M < N$). The system is said to be underdetermined because BSS cannot recover more than

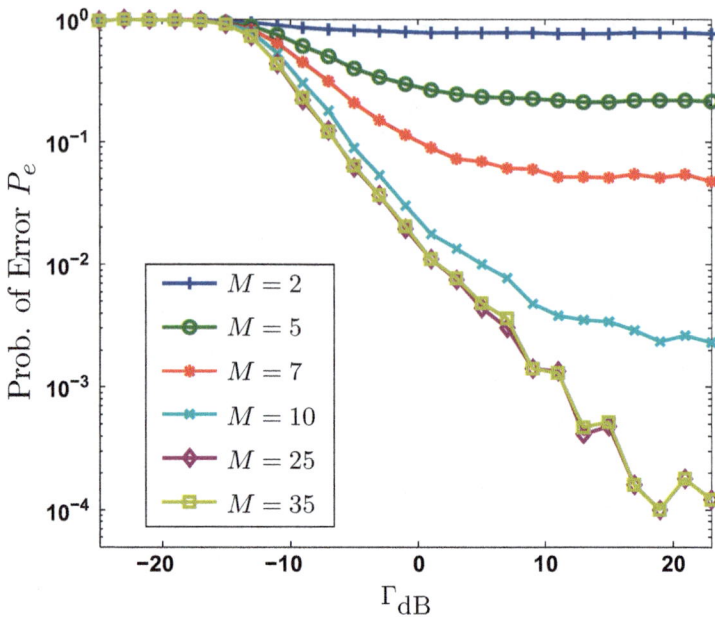

Fig. 7.7 Probability of error vs. Γ_{dB} with various Nb. obs. M. Based on PPP model: Avg. PD source density $\lambda_s = 4$ sources per unit volume or surface, avg. emission density $\lambda_e = 5$ discharges per unit time per sources

M sources. Nevertheless, there are at least M PD sources. When the number of observations is greater than the number of PD sources ($M > N$), the system is said to be overdetermined and the probability of error P_e decreases drastically because BSS can recover the exact number of PD sources. However, a very large number of observations does not provide a better performance.

A rapid identification of PD sources is proposed by which the number of sources can be estimated in power equipments for remote monitoring applications using a low cost Wireless IED. The technique is based on BSS via a generalized eigenvalue decomposition and the number of sources can be estimated by measuring the excess kurtosis. If we assume that the sources are non-white and decorrelated, or non-Gaussian and independent, the exact number of PD sources can be identified with a low probability of error, especially when discharges are significant. Performances can be affected by an underdetermined problem in BSS.

7.6 Conclusion

When EMIs from PD activity generate transient RF signals, it is not reasonable to reproduce such impulsive noise processes in substations via memoryless impulsive noise models because the resulting noise processes are *iid*, which implies that the PSD is constant over all frequencies. However, experimentation shows that, on average, the PSD of RF signals from PDs decays over frequency.

In this chapter, we have proposed a novel generalized impulsive noise model for substations. We have used a filtered Poisson process so that first- and second-order statistics can be derived based on practical and reasonable assumptions. The radio-noise process has been mathematically formulated based on the physics of PDs and the induced wave propagation. Taking advantage of the Poisson field of interferers model, some interesting statistical properties of moments, cumulants and probability distributions have been identified. These properties can be used for characterizing interference phenomena, performing communications analyses, and designing and optimizing communication systems or electronic intelligent devices for PD diagnostics in HV equipment.

The efficiency of the approach has been shown by comparing first- and second-order statistics of measurement and simulation results. The transient RF signals induced by PD sources are emulated by the time series models with first-order statistics of PDs obtained from data via a simple estimation procedure.

A practical application has been presented using the generalized impulsive noise model. We have developed a method for rapid identification of PD sources in HV equipment. It has been found that the number of PD sources can be estimated in arbitrary HV equipment via blind source separation using a generalized eigenvalue decomposition and the fourth-moment of the noise process.

Future work may focus on the comparison of impulsive noise models recently proposed by [140] using partitioned Markov chain models and/or any other existing impulsive noise models whose second-order statistics are taken into account as developed by [23, 99, 100]. The proposed rapid identification of PD sources can be tested experimentally in a laboratory with a definite number of PD sources. We believe that signal processing algorithms using the fourth-order moment as proposed by [191] can be implemented for PD localization together with the proposed PD identification methods.

Chapter 8
Conclusions

This monograph contributes to investigating the reliability of wireless sensor networks in substations by analysing, characterizing and modeling EMIs caused by PD sources. We have studied the electromagnetic noise that is generated in substation for a band that contains most of classic wireless carrier frequencies. The RF environment, called impulsive noise, is composed of series of impulses shaped as damped oscillations that are significant enough to be detected by existing wireless receivers. We have proposed two impulsive noise models. The first one is a Markov-Gaussian model configured as a partitioned Markov chain (PMC) that can generate a synthetic impulsive noise with a correlation between the samples that is more similar to the measured samples than with other models with memory. The PMC model can be implemented in a detector block of an optimal receiver that uses the memory of the model to mitigate the impact of impulsive noise. The second model is the first to propose a complete and coherent approach that links physical characteristics of high-voltage installations to the induced radio-interference spectrum. This allows for the design of more rapid and on-line partial discharge (PD) diagnostic tools when electromagnetic (EM) radiations are generated by PD sources in high voltage (HV) equipment, as well as the deployment of more robust wireless communication systems against man-made noise in the environment typical of substations.

The last chapter reviews the main highlights of the research, possible extensions, and the possible impact on research in the field of both PD diagnostic in HV equipment and communication in the presence of impulsive noise.

8.1 Monograph Summary

In Chap. 2, we have highlighted the necessity to improve the current models of impulsive noise for a wide band representation by considering the correlation between the samples. The literature background shows that wireless communica-

© Springer International Publishing AG, part of Springer Nature 2019   169
B. L. Agba et al., *Wireless Communications for Power Substations:*
RF Characterization and Modeling, Wireless Networks,
https://doi.org/10.1007/978-3-319-91328-5_8

tions are threatened by an highly correlated impulsive noise that requires models with memory in order to reproduce the electromagnetic environment of substation in a more accurate way than classic models.

In Chap. 3, we have described the setup and outcomes of an extensive measurement campaign. The measurement campaign has allowed to verify the damped oscillation shape of the impulses and the correlation between the equipment voltage and the power of the noise. The measurement setup and the results are shown in Chap. 3 with representative parameters that can be used by existing impulsive noise models. Using an RF measurement setup with a wideband antenna, we have captured signals from PD. A characterization procedure has been proposed for which first- and second-order statistics of PDs have been measured and characterized experimentally via advanced signal-processing tools. Our approach allows for the evaluation of the electromagnetic compatibility of PD interference sources with any electronic communication devices in the range of 800 MHz to 5 GHz.

It has been found that such spectral interferences can cover the frequency range of 800 MHz to 2 GHz on average. The RF signals from PD activity are characterized by very short rise times so that the spectra can reach 3 GHz. Thus, PDs are a major source of interferences for conventional wireless communications in ISM bands, in particular for the IEEE 802.15.4 standard which uses 915 MHz, 868 MHz and 2.4 GHz bands in the Americas, Europe and worldwide. Experimentations have demonstrated that IEEE 802.15.4 standard performs poorly in the presence of PDs and even worst when IEEE 802.11 standards are used in the 2.4 GHz bands [71, 72, 114, 118, 119]. As a result, substation environments can pose challenges to the reliability of wireless sensor networks due to PD phenomena.

One possible solution is to use an accurate and efficient wireless network deployment by positioning sensors in strategic locations to avoid EM radiations from PD activity. Our RF measurement setup can be used to assess the RF interference in substations. The level of the amplitude, number of impulses, and duration might be excellent criteria for assessing electromagnetic interference. Another possible solution is to use some communication systems that operate at higher frequency. For example, IEEE 802.11a/n/ac use 5 GHz bands, and in Canada, IEEE 802.16 standards use 3.5 or 5.8 GHz bands. To our knowledge, there is no research that investigates the impact of PD on the performance of such wireless communications. Our RF measurement setup can be extended for higher frequency bands by using an RF amplifier and a bandpass filter that operates over a large frequency range (i.e. 2–10 GHz).

At the same time, the frequency range of 800 MHz to 2 GHz can be used for PD measurement, detection, identification and localization, thereby allowing the development of rapid on-line remote monitoring and diagnostic tools in HV equipment. PD phenomena are a category of stochastic processes whose characteristics can be described as time-dependent random variables. With AC voltages, the PD process is especially cyclostationary as the PD occurs at every half-cycle of the applied voltage.

In Chap. 4, we proposed validated EMI model that links the discharge process to the induced far-field wave propagation. Assuming a PD source to be an electric dipole, the induced currents and charges have been linked to a magnetic potential vector source and an electric scalar potential source in which potentials can be expressed by solving Lorenz gauge condition equations in the far-field region. This allows for the derivation of derive the electromagnetic radiation of PD while keeping the cyclostationary process induced by AC voltages.

Using our proposed characterization process, we have successfully fulfilled validated the effectiveness of our approach. We used an LTI filter to reproduce the transient behaviour, and spectral characteristics were then estimated from data. Even though the LTI filter fits in terms of second-order statistics, the approach needs to be refined because distortions induced by multipath effects are not considered.

In Chap. 5, we generalized the approach by which spectral characteristics of PD and distortions can be estimated from data via a simple procedure. We have shown that time-series models are a natural generalization of LTI filter models, from which the estimation procedure and a measure of the goodness-of-fit are well-established in the literature [143–147]. The main issue is the selection of the time series model and the number of parameters to be estimated. There is a clear trade-off between the accuracy and the complexity of such models. Fortunately, this can be resolved by using the Akaike Information Criterion [146, 159] or the Schwarz Bayesian Information Criterion [147].

The goodness-of-fit has been measured by an analysis of the residuals from the fitted filters. We have shown that the effect of distortion can be modelled as a time-dependent Gaussian noise (i.e. heteroskedasticities in the residuals). These effects can be estimated via time series models with conditional heteroskedasticity such as ARCH, GARCH, or EGARCH. The analysis of the goodness-of-fit and a comparison between measurement and simulation results show the adequacy of the approach.

In Chap. 6, we have designed our own configuration of a partitioned Markov chain which can generate impulsive noise sample. The samples have an appropriate correlation that can observed in time domain as damped oscillating impulse. In frequency domain, the power spectrum of the synthetic impulses is very similar to the power spectrum calculated from the measurements, which other existing models cannot achieve. Such a model can be used to represent any substation environment for the 780 MHz to 2.5 GHz band; we have also provided representative parameters for 25, 230, 315 and 735 kV substation areas.

Chapter 7 focuses on extending the radio-noise model to contexts in which there are many PD sources that are randomly distributed in space. The Poisson field of interferers has been used because it allows for the identification of some interesting statistical properties of moments, cumulants and probability distributions that are linked by the physics of RF noise in substations. These can, in turn, be utilized in signal processing algorithms for rapid PD identification, localization, and impulsive noise mitigation techniques in wireless communications in substations.

Based on reasonable assumptions regarding the physical process of PD and the propagation of EM waves in the far-field region, first- and second-order statistics have been derived analytically by taking advantage of the Poisson field of inter-ferers. We have demonstrated that the probability distribution of the noise process can be approximated by an α-stable distribution. However, the main drawback of such an approximation is that first- and higher-order moments do not exist. In particular, the variance is infinite, which is not physically relevant because the power of an impulsive signal is finite. Under such conditions, authors in [183] have suggested using fractional lower-order moments for practical engineering applications. However, they have stated that these lower-order moments are much harder to work with than second- or higher-order moments because they introduce non-linearity to even linear problems. Fortunately, Campbell's theorem allows for the derivation of first- or higher-order moments in closed-form. These moments are finite if and only if first- and higher-order moments of the basic transient impulsive waveform and its random parameters are finite.

The effectiveness of the proposed model has been shown by comparing first- and second-order statistics of measurement and simulation results. The simulation parameters have been estimated from data using a simple estimation procedure.

As mentioned in many publications, the deployment of recent and advanced communications technology in substation environments is an area of growing interest [1–3, 7, 8]. This research has contributed to investigating the reliability of wireless sensor networks in substations by analysing, characterizing and modeling EMIs caused by PD sources. This allows for the design of more rapid and on-line PD diagnostic tools when EM radiations are generated by PD sources in HV equipment, as well as the deployment of more robust wireless communication systems against man-made noise in the environment typical of substations. We have successfully characterized EMIs induced by PD sources, and formalized a radio-noise model that links the discharge process to the induced far-field wave propagation. To our knowledge, we have first proposed a complete and coherent approach that the links physical characteristics of high-voltage installations to the induced radio-interference spectrum. The last chapter reviews the main highlights of the research, possible extensions, and the possible impact on research in the field of both PD diagnostic in HV equipment and communication in the presence of impulsive noise.

8.2   On the Practical Use of the EMI Models

We anticipate that several practical applications can be implemented for perfor-mance analyses, and the design and optimization of both PD diagnostic tools and wireless communication systems.

PD diagnostic tools in HV equipment: A new rapid identification of PD sources in HV equipment has been presented. Through an array of antennas, second-order statistics (the spatial covariance matrix) have been exploited in order to estimate the number of PD sources in arbitrary HV equipment. The generalized eigenvalue decomposition allows for the separation of PD sources. The latter can be counted by measuring the kurtosis (i.e. fourth-order of moment) at each antenna.

A performance analysis has been provided when the number of PD sources is not known by the receiver. It has been shown that the performance can be affected when the number of antennas is lower than the number of PD sources. The algorithm might be improved by estimating the number of strong PD sources in HV equipment. This can be done by measuring the kurtosis because a high value of the kurtosis suggests that the impulsive component is greater than the level of the background noise.

The literature has largely investigated PD localization problems in assessing insulation conditions in aged HV equipment. Several publications show that this is an area of intensive research [61, 64, 169, 170, 192, 193]. PD localization methods use the time difference of arrival (TDOA) between signals that are captured by an array of sensors or antennas. Although such methods show high accuracy, their effectiveness is based on: first, the assumption that there is a temporal correlation between signals captured by different sensors, which actually depends on the position of sensors [192]; second, that an expensive fast digital oscilloscope is needed to visualize a TDOA [61, 194]; and third, that the arrival time between sensors is determined based on human judgement [64, 193]. We believe that more rapid, cost-effective and automated PD localization algorithms can be developed based on spatial second- or higher-order statistics as summarized by [195, 196]. One of the most interesting methods is the use of spatial fourth-order moment, as proposed by [191], since background noise can be suppressed regardless of its coloring [196]. Such approaches are only based on spatial statistical assumptions of PD sources regardless of either temporal correlation or arrival time of impulses between sensors.

Wireless communication in substation environments: Performing communications analyses and designing and optimizing communication systems in such harsh and hostile environments can be done through our proposed generalized radio-noise model. Recently, we have proposed a novel impulsive noise mitigation technique with multi-antenna systems [197]. When PD sources are located in a specific region, those transient impulses arrive via the array of antennas with an arbitrary arrival angle. In narrowband, the interference signals received at each antenna differ only by amplitude and phase shifts. Taking a reference signal measured by a given sensor, the phase shift of the strongest impulse is estimated. This allows for impulsive noise mitigation, in which interference signals are cancelled successively. In addition, a new decision rule has been proposed to manage the effects of both background noise and impulsive noise components for decoding the transmitted message.

Simulation results show that the proposed receiver outperforms conventional receivers that are robust against Gaussian noise; this effect is amplified even more when the number of antennas is increased. However, the decision rule for impulsive noise mitigation is based on a heuristic weighting factor. This factor should be increased as the relative intensity of impulsive interferences is high; in other words, the degree of impulsiveness (the fourth-order of moment) is one important feature in the decision rule [197].

The mitigation of impulsive noise can be improved by working on both noise modeling and receiver design. We have worked on a wide band framework in order to consider many possibilities of wireless technology, which has as a consequence to model damped oscillations within the impulse shape. These radio-noise models can be used to plan and optimize wireless networks in substations via some interesting signal processing methods for impulsive noise mitigation techniques. In addition, new optimal and/or sub-optimal decision rules for decoding messages in harsh and hostile environments can by developed.

As future work, the wide band statistical model can be modified to provide a narrow band model, which requires a new configuration of the PMC with the appropriate parameters. From the power spectrum, we should be capable of predicting the amplitude and the durations of the impulses for specific band and bandwidth. The PMC model can be transformed into a narrow band model of impulsive noise by decreasing the number of states and by associating to each state a zero-mean Gaussian distribution.

The IED receivers must be autonomous, which requires adaptation to the substation environment by implementing a parameter estimation algorithm of the PMC model that is performed on a specific period. Also, it will be interesting to study a communication impaired by substation impulsive noise by using the representative parameters that we have calculated in this monograph work. Finally, a hardware implementation of the PMC model into a wireless receiver would be the most relevant objective to reach; it could be tested in high-voltage laboratory and in an existing substation.

References

1. Y. Yan, Y. Qian, H. Sharif, and D. Tipper, "A survey on smart grid communication infrastructures: Motivations, requirements and challenges," *IEEE Communications Surveys and Tutorials*, vol. 15 no 1, pp. 5–20, 2013.
2. V. C. Gungor, D. Sahin, T. Kocak, S. Ergut, C. Bucella, C. Cecatti, and G. P. Hancke, "Smart grid technologies : Communication technologies and standards," *IEEE Transactions on Industrial Informatics*, vol. 7 no 4, pp. 529–539, 2011.
3. S. M. Amin and B. F. Wollenberg, "Toward a smart grid: Power delivery for the 21st century," *IEEE Power and Energy Magazine*, vol. 3 no 5, pp. 34–41, 2005.
4. H. Farhangi, "The path of the smart grid," *IEEE Power and Energy Magazine*, vol. 8 no 1, pp. 18–29, 2010.
5. US Department of Energy Office, "A vision for the modern grid," National Energy Technology Laboratory, Tech. Rep., 2007.
6. ——, "A systems view of the modern grid," National Energy Technology Laboratory, Tech. Rep., 2007.
7. V. Gungor, B. Lu, and G. P. Hancke, "Opportunities and challenges of wireless sensor networks in smart grid," *IEEE Transactions on Industrial Electronics*, vol. 57 no 10, pp. 3557–3564, 2010.
8. V. Gungor and F. Lambert, "A survey on communication networks for electric system automation," *Computer Networks Elsevier*, vol. 50 no 7, pp. 877–897, 2006.
9. F. P. Sioshansi, *Smart grid, Integrating Renewable, Distributed, and Efficient Energy*. Academic Press, 2012.
10. C. Ozansoy, *Modelling and Object Oriented Implementation of IEC 61850*. Lambert Academy Publishing, 2010.
11. S. Chen, "A novel TD-LTE frame structure for heavy uplink traffic in smart grid," in *IEEE Innovative Smart Grid Technologies - Asia (ISGT Asia)*, 2014.
12. X. Yuzhe and C. Fischione, "Real-time scheduling in LTE for smart grids," in *5th International Symposium on Communications Control and Signal Processing (ISCCSP)*, 2012.
13. J. Du and M. Qian, "Research and application on LTE technology in smart grids," in *7th International ICST Conference on Communications and Networking in China (CHINACOM)*, 2012.
14. P. Parikh, T. Sidhu, and A. Shami, "A comprehensive investigation of wireless LAN for IEC 61850-based smart distribution substation applications," *IEEE Transactions on Industrial Informatics*, vol. 9, no. 3, pp. 1466–1476, 2013.

© Springer International Publishing AG, part of Springer Nature 2019
B. L. Agba et al., *Wireless Communications for Power Substations:
RF Characterization and Modeling*, Wireless Networks,
https://doi.org/10.1007/978-3-319-91328-5

15. G. GangJun, H. JunWei, Z. JuanXi, X. Chen, P. An, and W. XingChuan, "Design of routing optimization for substation high voltage side monitoring system based on wireless sensor network," in *International Conference on Advanced Power System Automation and Protection (APAP)*, 2011.

16. X. Long, Y. Yuan, and X. Jiang, "A novel low-power 802.11 wireless communication system for power substation applications," in *12th IEEE International Conference on Solid-State and Integrated Circuit Technology (ICSICT)*, 2014.

17. B. Bilgin and V. Gungor, "On the performance of multi-channel wireless sensor networks in smart grid environments," in *Proceedings of 20th International Conference on Computer Communications and Networks (ICCCN)*, 2011.

18. W. E. Pakala, E. R. Taylor, Jr., and R. T. Harrold, "Radio noise measurements on high voltage lines from 2.4 to 345 kV," in *IEEE Electromagnetic Compatibility Symposium Record*, 1968, pp. 96–107.

19. W. Pakala and V. Chartier, "Radio noise measurements on overhead power lines from 2.4 to 800 kV," *IEEE Transactions on Power Apparatus and Systems*, vol. PAS-90, issue 3, pp. 1155–1165, 1971.

20. C. H. Peck and P. Moore, "A direction-finding technique for wide-band impulsive noise source," *IEEE Transactions on Electromagnetic Compatibility*, vol. 43, no. 2, pp. 149–154, 2001.

21. I. Portugués, P. I. Moore, and I. A. Glover, "Characterization of radio frequency interference from high voltage electricity supply equipment," *ICAP. twelfth International Conference on Antennas and Propagation*, vol. 2, pp. p 820–823, 2003.

22. M. Hikita, H. Yamashita, T. Hoshino, T. Kato, N. Hayakawa, T.Ueda, and H. Okubo, "Electromagnetic noise spectrum caused by partial discharges in air at high voltage substations," *IEEE Transactions on Power Delivery*, vol. 13, No 2, pp. 434–439, 1998.

23. M. Zimmermann and K. Dostert, "Analysis and modelling of impulsive noise in broadband power-line communication," *IEEE Transactions on Electromagnetic Compatibility*, vol. 44 No 1, pp. 249–258, 2002.

24. D. Middleton, "Non-Gaussian noise models in signal processing for telecommunications: New methods and results for class A and class B noise models," *IEEE Transaction on Information Theory*, vol. 45 No 4, pp. 1129–1149, 1999.

25. ——, "Statistical-physical models of electromagnetic interference," *IEEE Transactions on Electromagnetic Compatibility*, vol. 19 Issue 3, pp. 106–127, 1977.

26. D. Fertonani and G. Colavolpe, "On reliable communications over channels impaired by bursty impulse noise," *IEEE transactions on Communications*, vol. 57 No 7, pp. 2024–2030, 2009.

27. K. L. Kaiser, *Electromagnetic Compatibility Handbook*, C. Press, Ed. CRC Press, 2005.

28. G. Vasilescu, *Electronic Noise and Interfering Signals: Principles and Applications*, Springer-Verlag, Ed. Springer, 2005.

29. V. Degardin, M. Lienard, A. Zeddam, F. Gauthier, and P. Degauquel, "Classification and characterization of impulsive noise on indoor power-line used for data communications," *IEEE Transactions on Consumer Electronics*, vol. 48 Issue:4, pp. 913–918, 2002.

30. CIGRÉ, *Perturbations Engendrées Par l'effet de Couronne des Réseaux de Transport : Description des Phénomènes Guide Practique de Calcul*. Conseil International des Grands Réseaux Électriques : Groupe de Travail 36.01 (Perturbations), 1974.

31. C. Gary, "Effet couronne sur les reseaux electriques aeriens," *Techniques de l'ingénieur. Génie électrique*, vol. D9, pp. D4440.1–D4440.25, 1998.

32. J. McDonald, *Electric Power Substations Engineering*. Leonard L. Grigsby, 2007.

33. C. Bayliss, *Transmission and distribution electrical engineering*. Oxford : Newnes, 2011.

34. R. Feinberg, *Modern Power Transformer Practice*. John Wiley and Sons, 1979.

35. R. Bartnikas and E. J. McMahon, *Engineering Dielectrics volume I Corona Measurement and Interpretation*, I. W. D. USA, Ed. American Society for Testing and Materials, 1979.

36. P. Chang, S. Chen, and J. He, "Lightning impulse corona characteristic of 1000-kV UHV transmission lines and its influences on lightning overvoltage analysis results," *IEEE Transactions on Power Delivery*, vol. 28, no. 4, pp. 2518–2525, 2013.

37. P. H. Moose and J. M. O'dwyer, "A model for impulsive power-line radio disturbance due to gap-type discharges," *IEEE Transactions on Electromagnetic Compatibility*, vol. 28 Issue:4, pp. 185–192, 1986, partial Discharge EM Noise.
38. A. Phillips, D. Childs, and H. Schneider, "Water drop corona effects on full-scale 500 kV non-ceramic insulators," *IEEE Transactions on Power Delivery*, vol. 14, no. 1, pp. 258–265, 1999.
39. T. Zhao and M. Comber, "Calculation of electric field and potential distribution along nonceramic insulators considering the effects of conductors and transmission towers," *IEEE Transactions on Power Delivery*, vol. 15, no. 1, pp. 313–318, 2000.
40. I. Portugues, P. Moore, I. Glover, C. Johnstone, R. McKosky, M. Goff, and L. van der Zel, "Rf-based partial discharge early warning system for air-insulated substations," *IEEE Transactions on Power Delivery*, vol. 24, no. 1, pp. 20–29, 2009.
41. M. D. Judd, L. Yang, and I. B. Frut, "Partial discharge monitoring of power transformers using UHF sensors. part I: Sensors and signal interpretation," *Electrical Insulation Magazine, IEEE*, vol. 21 Issue:2, pp. 5–14, 2005.
42. M. Pous and F. Silva, "Full-spectrum APD measurement of transient interferences in time domain," *IEEE Transactions on Electromagnetic Compatibility*, vol. 56, no. 6, pp. 1352–1360, 2014.
43. G. Bucci and C. Landi, "Measurement techniques for the characterization of wireless communication systems in time domain," in *Proceedings of the 18th IEEE Instrumentation and Measurement Technology Conference, IMTC*, 2001.
44. J. Khangosstar, Z. Li, and A. Mehboob, "IEEE international symposium on power line communications and its applications (ISPLC)," in *An experimental analysis in time and frequency domain of impulse noise over power lines*, 2011.
45. Q. Shan, I. Glover, R. C. Atkinson, S. A. Bhatti, I. E. Portugues, P. J. Moore, R. Rutherford, R. J. Watson, M. de Fatima Q. V., A. M. N. Lima, and B. A. Souza, "Estimation of impulsive noise in an electricity substation," *IEEE Transaction on Electromagnetic Compatibility*, vol. 53 No 3, pp. 653–663, 2011.
46. Q. Shan, I. Glover, R. Rutherford, S. Bhatti, R. Atkinson, I. Portugues, and P. Moore, "Detection of ultra wideband impulsive noise in a 400 kV air insulated electricity substation," in *20th International Conference and Exhibition on Electricity Distribution Part 1. CIRED*, 2009.
47. E. Kuffel, W. Zaengl, and J. Kuffel, *High Voltage Engineering : Fundamentals Second Edition*, N. E. Ltd., Ed. Elsevier Ltd., 2000.
48. L. B. Loeb, *Electrical Coronas: Their Basic Physical Mechanisms*, U. of California Press Berkeley and L. Angeles, Eds. University of California Press, 1965.
49. S. J. Townsend, *The Theory of Ionization of Gases by Collision*, C. London and Company, Eds. London, Constable and Company, ltd, 1910.
50. J. M. Meek and J. D. Craggs, *Electric Breakdown of Gases*, O. at the Clarendon Press, Ed. Oxford University Press, Amen House, London, 1953.
51. E. Lemke, "Guide for partial discharge measurements in compliance to IEC 60270," CIRGRE, Tech. Rep., 2008.
52. C. Hudon and M. Bélec, "Partial discharge signal interpretation for generator diagnostics," *IEEE Transactions on Dielectrics and Electrical Insulation*, vol. 12 No 2, pp. 297–319, 2005.
53. L. B. Loeb and J. M. Meek, "The mechanism of spark discharge in air at atmospheric pressure," *Journal of Applied Physics*, vol. 11, pp. 438–447, 1940.
54. R. J. V. Brunt and S. V. Kulkarni, "Stochastic properties of trichel-pulse corona: A non-Markovian random point process," *Physical Review. A, General Physics*, vol. 42 No 8, pp. 4908–4932, 1990.
55. R. J. V. Brunt, "Stochastic properties of partial-discharge phenomena," *IEEE Transactions on Electrical Insulation*, vol. 26 No 5, pp. 902–947, 1991.
56. M. Levesque, E. David, C. Hudon, and M. Belec, "Effect of surface degradation on slot partial discharge activity," *IEEE Transactions on Dielectrics and Electrical Insulation*, vol. 17 No 5, pp. 1428–1440, 2010.

57. V. L. Chartier, R. Sheridan, J. N. DiPlacido, and M. O. Loftness, "Electromagnetic inter-ference measurements at 900 MHz on 230 kV and 500 kV transmission lines," *IEEE Transactions on Power Delivery*, vol. 1 Issue:2, pp. 140–149, 1986.

58. K. Arai, W. Janischewskyj, and N. Miguchi, "Micro-gap discharge phenomena and television interference," *IEEE Transactions on Power Apparatus and Systems*, vol. PAS-104 No 1, pp. 221–232, 1985.

59. IEC-60270, "High-voltage test techniques - partial discharge measurements," IEC (Interna-tional Electrotechnical Commission), Tech. Rep., 2000.

60. S. Hoek, M. Koch, A. Kraetge, P. WInter, and M. HeInd, "Emission and propagation mechanisms of PD pulses for UHF and traditional electrical measurements," in *IEEE International Symposium on Electrical Insulation (ISEI), Conference Record*, 2012.

61. S. Tenbohlen, D. Denissov, S. M. Hoek, and S. M. Markalous, "Partial discharge measurement in the ultra high frequency (UHF) range," *IEEE Transactions on Dielectrics and Electrical Insulation*, vol. 15, No 6, pp. 1544–1553, 2008.

62. J. S. Pearson, B. F. Hampton, and A. G. Sellars, "A continuous uhf monitor for gas-insulated substations," *IEEE Transactions on Electrical Insulation*, vol. 26,no 3, pp. 469–478, 1991.

63. I. E. Portugués and P. J. Moore, "Study of propagation effects of wideband radiated RF signals from PD activity," in *IEEE Power Engineering Society General Meeting*, 2006.

64. P. J. Moore, I. E. Portugués, and I. A. Glover, "Radiometric location of partial discharges sources on energized high-voltage plant," *IEEE Transaction on Power Delivery*, vol. 20 no. 3, pp. 2264–2272, 2005.

65. T. Babnik, R. K. Aggarwal, P. J. Moore, and Z. D. Wang, "Radio frequency measurement of different discharges," *Power Tech. Conference Proceedings IEEE Bologna*, vol. 3, pp. 1–5, 2003.

66. G. N. Trinh, *The Electric Power Engineering Handbook*. CRC Press, 2001, ch. 5 Corona and Noise.

67. G. Madi, B. Vrigneau, Y. Pousset, R. Vauzelle, and B. L. Agba, "Impulsive noise of partial discharge and its impact on a minimum distance-based precoder of MIMO system," in *18th European Signal Processing Conference (EUSIPCO-2010)*, 2010, pp. 1602–1606.

68. L. Niemeyer, "A generalized approach to partial discharge modelling," *IEEE Transactions on Dielectrics and Electrical Insulation*, vol. 2 No 4, pp. 510–528, 1995.

69. F. Gutfleisch and L. Niemeyer, "Measurement and simulation of PD in epoxy voids," *IEEE Transactions on Dielectrics and Electrical Insulation*, vol. 2 No 5, pp. 729–743, 1995.

70. M. Levesque, E. David, and C. Hudon, "Effect of surface conditions on the electric field in air cavities," *IEEE Transactions on Electrical Insulation*, vol. 20 No 1, pp. 71–81, 2013.

71. S. A. Bhatti, Q. Shan, I. A. Glover, R. Atkinson, I. E. Portugues, P. J. Moore, and R. Rutherford, "Impulsive noise modelling and prediction of its impact on the performance of WLAN receiver," in *17th European Signal Processing Conference*, 2009, pp. 1680–1684.

72. G. Madi, F. Sacuto, B. Vrigneau, B. L. Agba, Y. Pousset, R. Vauzelle, and F. Gagnon, "Impacts of impulsive noise from partial discharges on wireless systems performance: Application to MIMO precoders," *EURASIP Journal on Wireless Communications and Networking*, vol. 186, pp. 1–12, 2011.

73. S. M. Zabin and H. V. Poor, "Efficient estimation of class A noise parameters via the EM [expectation-maximization] algorithms," *IEEE Transaction on Information Theory*, vol. 37 no 1, pp. 60–72, 1991.

74. ——, "Parameter estimation for Middleton class A interference processes," *IEEE Transac-tions on Communications*, vol. 37 no10, pp. 1042–1051, 1989.

75. G. A. Tsihrintzis and C. L. Nikias, "Fast estimation of the parameters of alpha-stable impulsive interference," *IEEE Transactions on Signal Processing*, vol. 44 Issue 6, pp. 1492–1503, 1996.

76. D. Middleton, "Canonical and quasi-canonical probability models of class A interference," *IEEE Transactions on Electromagnetic Compatibility*, vol. 25 Issue 2, pp. 76–106, 1983.

77. J. M. Chambers, C. L. Mallows, and B. W. Stuck, "A method for simulating stable random variables," *Journal of the American Statistical Association*, vol. 71 No 354, pp. 340–344, 1976.

78. S. Kullback and R. A. Leiber, "On information and sufficiency," *The annals of Mathematical Statistics*, vol. 22 No 1, pp. 79–86, 1951.

79. F. J. Massey, "The Kolmogorov-Smirnov test for goodness of fit," *Journal of the American Statistical Association*, vol. 46, no. 253, pp. 68–78, 1951.

80. S. Mallat, *A Wavelet tour of Signal Processing*, Elsevier, Ed. Academic Press, 1998.

81. J. Hammond and P. White, "The analysis of non-stationary signals using time-frequency methods," *Journal of Sound and Vibration Elsevier*, vol. 190, Issue 3, pp. 419–447, 1996.

82. M. Priestley, "Power spectral analysis of non-stationary random processes," *Journal of Sound and Vibration Elsevier*, vol. 6, issue 1, pp. 86–97, 1967.

83. G. Ndo, F. Labeau, and M. Kassouf, "A Markov-Middleton model for bursty impulsive noise : Modelling and receiver design," *IEEE Transactions on Power Delivery*, vol. 28 No 4, pp. 2317–2325, 2013.

84. S. V. Vaseghi, *Advanced Digital Signal Processing and Noise Reduction : Fourth Edition*, J. W. . Sons, Ed. Wiley, 2008.

85. H. Ferreira, L. Lampe, J. Newbury, and T. Swart, *Power Line Communications: Theory and Applications for narrowband and broadband communications over power lines*. John Wiley and John Wiley & Sons Inc., 2010.

86. T. Li, W. H. Mow, and M. Siu, "Joint erasure marking and Viterbi decoding algorithm for unknown impulsive noise channels," *IEEE Transactions on Wireless Communications*, vol. 7, no. 9, pp. 3407–3416, 2008.

87. H. Vu, N. Tran, T. Nguyen, and S. Hariharan, "Estimating Shannon and constrained capacities of Bernoulli-Gaussian impulsive noise channels in Rayleigh fading," *IEEE Transactions on Communications*, vol. 62, no. 6, pp. 1845–1856, 2014.

88. ——, "On the capacity of Bernoulli-Gaussian impulsive noise channels in Rayleigh fading," in *International Symposium on Personal Indoor and Mobile Radio Communications (PIMRC)*, 2013.

89. A. Dubey, R. Mallik, and R. Schober, "Performance of a PLC system in impulsive noise with selection combining," in *IEEE Global Communications conference (GLOBECOM)*, 2012.

90. P. A. Delaney, "Signal detection in multivariate class-A interference," *IEEE Transactions on Communications*, vol. 43, no. 234, pp. 365–373, 1995.

91. D. Middleton, "Procedures for determining the parameters of the first-order canonical models of class A and class B electromagnetic interference," *IEEE Transactions on Electromagnetic Compatibility*, vol. 21 Issue 3, pp. 190–208, 1979.

92. C. Nikias and M. Shao, *Signal Processing with Alpha-Stable Distributions and Applications*, i. Wiley, Ed. Wiley, 1995.

93. P. Lévy, *Calcul des Probabilités*, G.-V. editor, Ed. Gauthier-Villars, 1925.

94. G. A. Tsihrintzis and C. L. Nikias, "Performance of optimum and sub-optimum receivers in the presence of impulsive noise modeled as an alpha-stable process," *IEEE Transactions on Communications*, vol. 43 Issue 234, pp. 904–914, 1995.

95. A. Spaulding and D. Middleton, "Optimum reception in an impulsive interference environment-part I: Coherent detection," *IEEE Transactions on Communications*, vol. 25, pp. 910–923, 1977.

96. K. C. Wiklundh, P. F. Stenumgaard, and H. M. Tullberg, "Channel capacity of Middleton's class A interference channel," *Electronics Letters*, vol. 45 no 24, pp. 1–2, 2009.

97. J. Wang, E. E. Kuruoglu, and T. Zhou, "Alpha-stable channel capacity," *IEEE Communications Letters*, vol. 15 no 10, pp. 1107–1109, 2011.

98. P. J. Moore, I. E. Portugués, and I. A. Glover, "Partial discharge investigation of a power transformer using wireless wideband radio-frequency measurements," *IEEE Transactions on Power Delivery*, vol. 21 No 1, pp. 528–530, 2006.

99. E. N. Gilbert, "Capacity of burst-noise channels," *Bell Syst. Tech. Journal*, vol. 39, pp. 1253–1266, 1960.

100. E. O. Elliot, "Estimates of error rates for codes on burst-noise channels," *Bell Syst. Tech. Journal*, vol. 42, pp. 1977–1997, 1963.

101. J. Mitra and L. Lampe, "Convolutionally coded transmission over Markov-Gaussian channels: Analysis and decoding metrics," *IEEE Transactions on Communications*, vol. 58, no. 7, pp. 1939–1949, 2010.

102. Q. Shan, I. A. Glover, P. J. Moore, I. E. Portugues, M. Judd, R. Rutherford, and R. J. Watson, "Performance of zigbee in electricity supply substations," in *WiCOM. International Conference on Wireless Communications, Networking and Mobile Computing*, 2007, pp. 3871–3874.

103. Y. Ma, P. So, and E. Gunawan, "Performance analysis of OFDM systems for broadband power line communications under impulsive noise and multipath effects," *IEEE Transaction on Power Delivery*, vol. 20 no 2, pp. 674–682, 2005.

104. J. Haring and H. Vinck, "OFDM transmission corrupted by impulsive noise," in *International Symposium on Power-Line Communications and its Applications*, 2000, pp. 5–7.

105. M. Ghosh, "Analysis of the effect of impulse noise on multicarrier and single carrier QAM systems," *IEEE Transaction on Communications*, vol. 44 no 2, pp. 145–147, 1996.

106. J. Ling, M. Nassar, and B. L. Evans, "Impulsive noise mitigation in power line communications using sparse Bayesian learning," *IEEE Journal Selected Areas in Communications*, vol. 31 no 7, pp. 1172–1183, 2013.

107. S. Ambike and J. J. L. Hatzinakos, "Detection for binary transmission in a mixture of Gaussian noise and impulsive noise modeled as an alpha-stable process," *IEEE Signal Processing Letters*, vol. 1 no 3, pp. 55–57, 1994.

108. J. G. Proakis and M. Salehi, *Digital Communications Fifth Edition*, M.-H. H. Education, Ed. McGraw-Hill Higher Education, 2008.

109. A. Goldsmith, *Wireless Communications*, C. U. Press, Ed. Cambridge University Press, 2005.

110. F. Cleveland, "Use of wireless data communications in power system operations," in *IEEE Power Engineering Society, Power System Communications Committee (PSCC) and Substation Committee (PSSC)*, 2006, pp. 1–10.

111. D. Moongilan, "Corona noise considerations for smart grid wireless communication and control network planning," in *IEEE International Symposium on Electromagnetic Compatibility (EMC)*, 2012.

112. B. Li, W. Song, and Y. Li, "The influence of strong electromagnetism field to wireless communication system," in *4th International Conference on Wireless Communications, Networking and Mobile Computing, WiCOM*, 2008.

113. A. Bo, Z. Weidong, C. Xiang, L. Jikun, S. Yingbin, and L. Shaoyu, "A study on immunity of wireless sensor unit in substation," in *International Symposium on Electromagnetic Compatibility (EMC EUROPE)*, 2012.

114. Q. Shan, I. Glover, P. J. Moore, I. E. Portugues, R. J. Watson, R. Rutherford, R. Atkinson, and S. Bhatti, "Laboratory assessment of WLAN performance degradation in the presence of impulsive noise," in *Wireless Communications and Mobile Computing Conference*, 2008, pp. 859–863.

115. F. Sacuto, "Adaptation d'un réseau sans fil de capteurs à une zone soumise à des interférences électromagnétiques dues aux hautes tensions," Master's thesis, École de technologie Supérieure Université Du Québec, 2010.

116. J. D. McDonald, *Electric Power Substations Engineering: Third Edition*. Cambridge University Press, 2012.

117. D. B. Percival and A. T. Walden, *Wavelet methods for time series analysis*. Cambridge University Press, 2000.

118. F. Sacuto, B. L. Agba, F. Gagnon, and F. Labeau, "Evolution of the RF characteristics of the impulsive noise in high voltage environment," in *IEEE SmartGridComm Workshop*, 2012.

119. S. A. Bhatti, Q. Shan, I. A. Glover, and R. Atkinson, "Performance simulation of WLAN and zigbee in electricity substation impulsive noise environments," in *IEEE SmartGridComm Workshop*, 2012.

120. D. Donoho and I. M. Johnstone, "Ideal spatial adaptation by wavelet shrinkage," *Biometrika*, vol. 81 No 3, pp. 425–455, 1994.

121. H. Krim, D. Tucker, S. Mallat, and D. Donoho, "On denoising and the best signal representation," *IEEE Transactions on Information Theory*, vol. 45 No 7, pp. 2225–2238, 1999.
122. X. Ma, C. Zhou, and I. J. Kemp, "Interpretation of wavelet analysis and its application in partial discharge detection," *IEEE Transactions on Dielectrics and Electrical Insulation*, vol. 9 No 3, pp. 446–457, 2002.
123. L. Satish and B. Nazneen, "Wavelet-based denoising of partial discharge signals buried in excessive noise and interference," *IEEE transactions on Dielectrics and Electrical Insulation*, vol. 10 No 2, pp. 354–367, 2003.
124. M. Au, F. Gagnon, and B. L. Agba, "An experimental characterization of substation impulsive noise for a RF channel model," *Progress In Electromagnetics Research Symposium, PIERS Proceedings*, vol. 1, pp. 1371–1376, 2013.
125. R. Bartnikas, "Partial discharges their mechanism, detection and measurement," *IEEE Transactions on Dielectrics and Electrical Insulation*, vol. 9 No 5, pp. 763–808, 2002.
126. C. Hudon, M. Bélec, and M. Lévesque, "Study of slot partial discharges in air-cooled generators," *IEEE Transactions on Dielectrics and Electrical Insulation*, vol. 15 No 6, pp. 1675–1690, 2008.
127. M. B. Wilk and R. Gnanadesikan, "Probability plotting methods for the analysis for the analysis of data," *Biometrika*, vol. 55 No. 1, pp. 1–17, 1968.
128. B. Fruth and L. Niemeyer, "The importance of statistical characteristics of partial discharge data," *IEEE Transactions on Electrical Insulation*, vol. 27 No 1, pp. 60–69, 1992.
129. V. Kogan, F. Dawson, G. Gao, and B. NIndra, "Surface corona suppression in high voltage stator winding end turns," *Electrical Electronics Insulation Conference and Electrical Manufacturing & Coil Winding Conference. Proceedings*, vol. 1. No 1, pp. 411–415, 1995.
130. E. David and L. Lamarre, "Low-frequency dielectric response of epoxy-mica insulated generator bars during multi-stress aging," *IEEE Transactions on Dielectrics and Electrical Insulation*, vol. 14 no 1, pp. 212–226, 2006.
131. I. Gallimberti, G. Marchesi, and R. Turri, "Corona formation and propagation in weakly and strongly attaching gases," *Proc. 8th Int. Conf. on Gas Discharges*, vol. 1 No 1, pp. 587–594, 1985.
132. M. von Laue, "Comment on K. Zuber's measurement of the spark discharge delay time," *Annals of Physics Liepzig*, vol. 76, pp. 261–265, 1925.
133. J. S. Chang, P. A. Lawless, and T. Yamamoto, "Corona discharge processes," *IEEE Transactions on Plasma Science*, vol. 19 No 6, pp. 1152–1166, 1991.
134. M. M. Rao, M. J. Thomas, and B. P. Singh, "Electromagnetic field emission from gas-to-air bushing in a GIS during switching operations," *IEEE Transactions on Electromagnetic Compatibility*, vol. 49, No 2, pp. 313–321, 2007, electrical arc.
135. T. Okazaki, Z. Kawasaki, and A. Hirata, "Wideband characteristics of impulsive EM noise emitted from discharges and development of mathematical noise model," in *EMC, International Symposium on Electromagnetic Compatibility*, vol. 2, 2005, pp. p 469–472.
136. S. Minegishi, H. Echigo, and R. Sato, "Frequency spectra of the arc current due to opening electric contacts in air," *IEEE Transactions on Electromagnetic Compatibility*, vol. 31, No 4, pp. 342–345, 1989, arc Electrique.
137. M. P. Shinde and S. N. Gupta, "A model of HF impulsive atmospheric noise," *IEEE Transaction on Electromagnetic Compatibility*, vol. EMC-16, No 2, pp. 71–75, 1974.
138. A. Weron and R. Weron, "Computer simulation of levy alpha-stable variables and processes," *Lecture Notes in Physics*, vol. 457, pp. 379–392, 1995.
139. G. Samarodnitsky and M. S. Taqqu, *Stable Non-Gaussian Random Processes : Stochastic Models with Infinite Variance*, C. P. LLC, Ed. CRC Press LLC, 2000.
140. F. Sacuto, F. Labeau, and B. L. Agba, "Wideband time-correlated model for wireless communications under impulsive within power substations," *IEEE Transaction on Wireless Communications*, vol. 13 No. 3, pp. 1449–1461, 2013.
141. M. Au, B. L. Agba, and F. Gagnon, "A model of electromagnetic interferences induced by corona discharges for wireless channels in substation environments," *IEEE Transactions on Electromagnetic Compatibility*, vol. 57, no 3, pp. 522–531, 2015.

142. B. M. Sadler, "Detection in correlated impulsive noise using fourth-order cumulants," *IEEE Transaction on Signal Processing*, vol. 14 No 11, pp. 2793–2800, 1996.

143. D. A. Dickey and W. A. Fuller, "Distribution of the estimators for autoregressive time series with a unit root," *Journal of the American Statistical Association*, vol. 74, No. 366, pp. 427–431, 1979.

144. P. B. Phillips and P. Perron, "Testing for a unit root in time series regression," *Biometrika*, vol. 75, No 2, pp. 335–346, 1988.

145. G. E. P. Box, G. M. Jenkins, and G. C. Reinsel, *Time Series Analysis: Forecasting and Control 3rd Ed.*, N. P. Hall, Ed. Englewood Cliffs, 1994.

146. H. Akaike, "Information theory and an extension of the maximum likelihood principle," in *Second International Symposium on Information Theory*, 1973, pp. 267–281.

147. G. Schwarz, "Estimating the dimension of a model," *The annals of Statistics*, vol. 6, No. 2, pp. 461–464, 1978.

148. M. Judd, B. Hampton, and O. Farish, "Modelling partial discharge excitation of UHF signals in waveguide structures using green's functions," *IEE Proceedings Science, Measurement and Technology*, vol. 143 No 1, pp. 63–70, 1996.

149. W. A. Woodward, H. L. Gray, and A. C. Elliott, *Applied Time Series Analysis*, C. P. Taylor and F. Group, Eds. CRC Press Taylor and Francis Group, 2012.

150. W. A. Fuller, *Introduction to Statistical Time Series. 2nd Ed.*, N. Y. Wiley, Ed. Wiley, 1996.

151. J. D. Hamilton, *Time Series Analysis*, P. U. Press, Ed. minus 0.4emPrinceton University Press, 1994.

152. T. Bollerslev, "Generalized autoregressive conditional heteroscedasticity," *Journal of Econometrics*, vol. 31, pp. 307–327, 1986.

153. R. F. Engle, "Autoregressive conditional heteroscedasticity with estimates of the variance of united kingdom inflation," *Journal of Econometrics*, vol. 50 No 4, pp. 987–1007, 1982.

154. D. Nelson, "Conditional heteroskedasticity in asset returns: A new approach," *Econometrica*, vol. 59, No. 2, pp. 347–370, 1991.

155. D. Straumann, *Estimation in Conditionally Heteroskedastic Time Series Models*, Springer, Ed. Springer, 2005.

156. S. Said and D. A. Dickey, "Testing for unit roots in autoregressive moving-average models of unknown order," *Biometrika*, vol. 71, pp. 599–607, 1984.

157. J. G. MacKinnon, "Approximate asymptotic distribution functions for unit-root and co-integration tests," *Journal of Business & Economic Statistics*, vol. 12, No. 2, pp. 167–176, 1994.

158. W. K. Newey and K. D. West, "A simple, positive semi-definite, heteroskedasticity and autocorrelation consistent covariance matrix," *Econometrica*, vol. 55 No. 3, pp. 703–708, 1987.

159. H. Akaike, "A new look at the statistical model identification," *IEEE Transaction on Automatic Control*, vol. 19 No. 6, pp. 716–723, 1974.

160. G. M. Ljung and G. E. P. Box, "On a measure of a lack of fit in time series models," *Biometrika*, vol. 65, issue 2, pp. 297–303, 1978.

161. G. E. P. Box and D. A. Pierce, "Distribution of residual autocorrelations in autoregressive-integrated moving average time series models," *Journal of the American Statistical Association*, vol. 65 No 332, pp. 1509–1526, 1970.

162. J. M. Chambers, W. S. Cleveland, B. Kleiner, and P. A. Tukey, *Graphical Methods for Data analysis*, Taylor and F. Group, Eds. CRC Press, 1983.

163. C. Hurvich and C. Tsai, "Regression and time series model selection in small samples," *Biometrika*, vol. 76, no. 2, pp. 297–307, 1988.

164. A. I. McLeod and W. K. Li, "Diagnostic checking ARMA time series models using squared-residual autocorrelations," *Journal of Time Series Analysis*, vol. 4. No. 4, pp. 269–273, 1983.

165. C. Granger and A. Andersen, *An Introduction of Bilinear Time Series Models*, Gottingen, Ed. Vandenhoeck and Ruprecht, 1978.

166. F. Sacuto, F. Labeau, J. Béland, M. Kassouf, S. Morin, S. Poirier, and B. Agba, "Impulsive noise measurement in power substations for channel modeling in ISM band," in *CIGRE Canada Conference*, 2012.

167. G. Gan, C. Ma, and J. Wu, *Data Clustering Theory, Algorithms, and Applications*. SIAM, 2007.
168. J. Kemeny and J. Snell, *Finite Markov Chains*. Springer-Verlag, 1976.
169. G. C. Montanari and A. Cavallini, "Partial discharge diagnostics: From apparatus monitoring to smart grid assessment," *IEEE Electrical Insulation Magazine*, vol. 29, no. 3, pp. 8–17, 2015.
170. M. Wu, H. Cao, J. Cao, H. L. Nguyen, J. B. Gomes, and S. P. Krishnaswamy, "An overview of state-of-the-art partial discharge analysis techniques for condition monitoring," *IEEE Electrical Insulation Magazine*, vol. 31, no. 6, pp. 22–35, 2015.
171. H. Ma, K. Saha, C. Ekanayake, and D. Martin, "Smart transformer for smart grid-intelligent framework and techniques for power transformer asset management," *IEEE Transaction on Smart Grid*, vol. 6, no. 2, pp. 1026–1034, 2015.
172. M. Au, B. L. Agba, and F. Gagnon, "A fast identification of partial discharge sources using blind source separation and kurtosis," *Electronic Letters IET*, vol. 1, no. 1, pp. 1–2, 2015, (accepted).
173. R. Bartnikas and J. P. Novak, "On the character of different forms of PD and their related terminologies," *IEEE Transactions on Electrical Insulation*, vol. 28, pp. 956–968, 1993.
174. D. Middleton, "Statistical-physical model of man-made radio noise, part I : First-order probability models of the instantaneous amplitude," United States Department of Commerce office of telecommunications, Tech. Rep., 1974.
175. ——, "Man-made noise in urban environments and transportation system : Models and measurements," *IEEE Transactions on Communications*, vol. 21, No 11, pp. 1232–1241, 1973.
176. E. Parzen, *Stochastic Processes*, S. S. for Industrial and A. Mathematics), Eds. SIAM, 1962.
177. M. Girault, *Stochastic Processes*, Springer-Verlag, Ed. Springer-Verlag, 1966.
178. D. L. Snyder and M. I. Miller, *Random Point Processes in Time and Space: Second Edition*, Springer-Verlag, Ed. Springer-Verlag, 1991.
179. S. O. Rice, "Mathematical analysis of random noise," *Bell Syst. Tech. Journal*, vol. 23, pp. 282–332, 1944.
180. ——, "Mathematical analysis of random noise : Part III," *Bell Syst. Tech. Journal*, vol. 24, pp. 46–156, 1945.
181. I. Gradshteyn and I. Ryzhik, *Tables of Integrals, Series and Products: Seventh Edition*, A. Jeffrey and D. Zwillinger, Eds. Academic Press, 2007.
182. I. A. Koutrouvelis, "An iterative procedure for the estimation of the parameters of stable laws," *Communications in Statistics-Simulation and Computation*, vol. 10 No 1, pp. 17–28, 1981.
183. M. Shao and C. Nikias, "Signal processing with fractional lower order moments: Stable processes and their applications," *Proceedings of the IEEE*, vol. 81, no. 7, pp. 986–1010, 1993.
184. S. B. Lowen and M. C. Teich, "Power-law shot noise," *IEEE Transactions on Information Theory*, vol. 36 No 6, pp. 1302–1317, 1990.
185. J. McCulloch, "Simple consistent estimators of stable distribution parameters," *Communications in Statistics - Simulation and Computation*, vol. 15, no. 4, pp. 1109–1136, 1986.
186. J. R. Carson, "The statistical energy-frequency spectrum of random disturbances," *Bell Syst. tech. Journal*, vol. 10 No 3, pp. 374–381, 1931.
187. L. Parra and P. Sajda, "Blind source separation via generalized eigenvalue decomposition," *Journal of Machine Learning Research*, vol. 4, pp. 1261–1269, 2003.
188. L. Parra and C. Spence, "Convolutive blind separation of non-stationary sources," *IEEE Transaction on Speech and Audio Processing*, vol. 8 no 3, pp. 320–327, 2000.
189. E. Weinstein, M. Feder, and A. V. Oppenheim, "Multi-channel signal separation by decorrelation," *IEEE Transaction on Speech and Audio Processing*, vol. 1 no 4, pp. 405–413, 1993.
190. J. F. Cardoso and A. Souloumiac, "Blind beamforming for non-gaussian signals," *IEE Proceedings F (Radar and Signal Processing)*, vol. 140 no 6, pp. 362–370, 1993.

191. E. Moulines and J.-F. Cardoso, "Asymptotic performance analysis of direction-finding algorithms based on fourth-order cumulant," *IEEE Transaction on Signal Processing*, vol. 57, no. 8, pp. 214–224, 1995.

192. H. Sinaga, B. Phung, and T. Blackburn, "Partial discharge localization in transformers using UHF detection method," *IEEE Transaction on Dielectrics and Electrical Insulation*, vol. 19, no. 6, pp. 1891–1900, 2012.

193. S. Markalous, S. Tenbohlen, and K. Feser, "Detection and location of partial discharges in power transformers using acoustic and electromagnetic signals," *IEEE Transaction on Dielectrics and Electrical Insulation*, vol. 15, no. 6, pp. 1576–1583, 2008.

194. Z. Tang, C. Li, X. Cheng, W. Wang, J. Li, and J. Li, "Partial discharge location in power transformers using wideband RF detection," *IEEE Transactions on Dielectrics and Electrical Insulation*, vol. 13, no. 6, pp. 1193–1199, 2006.

195. H. L. V. Trees, *Optimum Array Processing Part IV : Detection, Estimation and Modulation Theory*, J. Wiley and Sons, Eds. Wiley-Interscience, 2004.

196. E. Gonen and J. M. Mendel, *The Digital Signal Processing Handbook: Second Edition Wireless, NetNetwork, Sensor Array Processing and Nonlinear Signal Processing*, V. K. Madisetti, Ed. CRC Press Taylor and Francis Group, 2010.

197. I. B. S. Ali, M. Au, B. L. Agba, and F. Gagnon, "Mitigation of impulsive interference in power substation with multi-antenna systems," in *IEEE ICUWB*, 2015.

Index

© Springer International Publishing AG, part of Springer Nature 2019
B. L. Agba et al., *Wireless Communications for Power Substations:
RF Characterization and Modeling*, Wireless Networks,
https://doi.org/10.1007/978-3-319-91328-5

Printed by Printforce, the Netherlands